COMMON GROUND

Garden histories of Aotearoa

COMMON GROUND

Garden histories of Aotearoa

MATT MORRIS

OTAGO UNIVERSITY PRESS
Te Whare Tā o Te Wānanga o Ōtākou

Published by Otago University Press
533 Castle Street
Dunedin, New Zealand
university.press@otago.ac.nz
www.otago.ac.nz/press

First published 2020
Copyright © Matt Morris

The moral rights of the author have been asserted

ISBN 978-1-98-859257-2

Editor: Imogen Coxhead

Indexer: Lee Slater

Cover: © Shutterstock; Gavin Paterson (bending) with his daughter Eunice in the vegetable garden, Dannevirke, c. 1943. Photo courtesy of Lachy Paterson

Author photograph: Courtesy University of Canterbury

Designed by Smartwork Creative Ltd, www.smartworkcreative.co.nz

Printed in China through Asia Pacific Offset

Contents

Acknowledgements

This book has taken a long time to grow, and many have helped to nurture it along the way.

I wish to thank the people who taught me to love gardens, shared their gardening stories with me, or helped me to learn about New Zealand's history, garden history and the history of empire; and the dreamers who are implementing their vision of an equitable and sustainable future in which gardens will play a vital role.

Various people have helped to shape this text. Early encouragement for this project, which started life as a PhD, came from Brendan Hoare, a longtime leader in the organic movement, and from his mentor, Bob Crowder. They helped me identify ways in which gardening could be a *political* act. Numerous discussions with friends and colleagues within gardening circles have rested on this point. Among these many people are Bailey Peryman, Peter Wells, Rowan Brookes, David Border-Giles, Doug Hesp, Moko Morris, Marion Thomson, Gordyn Hamblyn, Steffan Browning, Clare Abaffey and Ami Kennedy.

I have been privileged to talk to many gardeners, especially elderly ones, about their gardens and those of their parents or grandparents. From these delightful conversations I learned an enormous amount about the role of family and neighbourhood networks in honing gardening skills and knowledge. I visited their wonderful gardens too, in all their variation and colour. The garden of Mrs Paraskevi, a Cypriot who had lived in New Zealand for seven decades, was crammed with roses and rosemary, and with vegetables and herbs grown from seed originally smuggled into the country after a trip 'home' in the 1970s: a slice of Cyprus in suburban Wellington.

Interviews with people from a single street or neighbourhood showed me how diverse and yet how connected gardening communities could be. The Clifton Hill residents stand out. Fern Every told me about her expansive vegetable garden, which she started in 1949, in a way recreating the hillside peasant gardens she grew up among in the mountains west of Beijing. But the gardens of Clifton were just as likely to be planted in fruit trees, New Zealand natives, South African plants, or crammed with cacti. The people I interviewed enriched my understanding of gardening, and I am extremely grateful to them all.

Several historians offered me important support and guidance in shaping this book. John Cookson and Philippa Mein Smith gave years of time to supervising my original thesis, while Geoff Rice provided valuable feedback on an earlier draft of this book. Angela Wanhalla's encouragement meant a great deal to me, and James Beattie's detailed reading of a later draft added immensely to the final version. I offer my heartfelt thanks to them, and to the garden historians who gave me time to tease out ideas early in the process: Helen Leach, Katie Holmes, Michel Conan, Joachim Wolschke-Buhlmann and Catherine Benoît.

I also thank those in the 'community food' and sustainability sectors that I get to work alongside, who constantly shape and reshape my thinking about the place of gardens in a world in crisis. I received a generous donation towards the project from the Canterbury branch of Soil & Health, and much inspiration from the national Soil & Health council, the Food Resilience Network, and the multitude of community gardeners and food foresters up and down the country. I'm glad there are so many of you.

I am extremely grateful to the team at Otago University Press for making this book possible, and especially to Imogen Coxhead for her careful editing of the text.

My partner, family and friends have journeyed with this book project for more than ten years. Thank you for your patience and support.

Prologue

Overgrown Christchurch

In December 2012 I joined two of my scholarship students on a field trip into Christchurch's residential red zone (now the Ōtākaro Avon River Corridor), the public exclusion zone implemented after the earthquake of 22 February 2011. The students and I were taking soil samples; we were feeding into a body of work seeking to establish new food gardens throughout the area that could underpin a resilient local food economy. Here, many of the previously privately-owned homes and gardens had been destroyed in the Christchurch earthquakes and subsequently purchased by the Crown. Some had long gardening histories, stretching back almost to 1850.

Between that old history and these new visions stood the gardens as they were. Many of the trees were intact, although others had been bulldozed in the first flush of clearances. Some gardens had become fields of poppies and eschscholzia. Through cracked driveways tall grasses were sending up impressive seed heads; lawns were now meadows.

The area we were in was generally close to the Ōtākaro Avon River. On River Road we called in to see Diana Madgin, a *Press* gardening columnist and local activist who was keen to see the gardens of the area preserved for historical, community and tourism purposes. In her recitation of some of the early colonial gardening history of Christchurch, she reminded me that my doctoral thesis on that topic, which she acknowledged as a key source of information for her, might actually be of some use to the wider community.

Where Diana lived had been an important orcharding area in the nineteenth century; further along the river there had been significant market gardens. And, of course, back in the other direction, at the place called Ōtautahi, Māori food-gathering connections stretched back into the seventeenth century if not earlier.

Gardeners understand why garden histories are important. When we talk about what is in a garden, we need to know what preceded it, because gardens are never made in a void. They are always shaped by what is already there. This is true even when we imagine that the area for a new garden is empty of garden references.

When I went wandering through the overgrown, abandoned gardens with my students, I felt a sudden sense of urgency about retelling the story of our gardens. What remains on those sections is an incredible legacy and an extraordinary opportunity, and the stories we have created over the decades deserve to be acknowledged, both individually and corporately. They have made us who we are as a community.

I was filled with joy – which I'm not sure my students really understood – when I discovered on some properties that lemon trees grown right up close to the houses had been retained even when the houses had been completely demolished. Grapevines along fences had likewise been granted a reprieve, even when the fences themselves had been removed. It was subtle, unspoken, but this gesture said to me in no uncertain terms that the great hope for expanded food growing in this area was considered a distinct possibility by those who make decisions. All of this underscored the feeling I was getting that the gardening histories I've been collecting for years deserve a proper airing, at this critical juncture in Christchurch's evolution.

Most of the research I've done has been about Christchurch, the Garden City of New Zealand. I've lived in Christchurch all of my life, I've been involved with many gardening organisations here, I have been an elected member of local government, and I even wrote a *Press* gardening column for several years. Christchurch is my home.

But I have also trawled through archives all over the country and beyond, and I've spoken with gardeners of all ages up and down both of our main islands. So while my focus is necessarily on Christchurch,

I feel compelled to weave those other stories from around Aotearoa into this narrative as well. I do this because New Zealand is a small country, and it seems to me that all of our stories intertwine beautifully. Furthermore, like gardens, colonial cities are never simply founded on top of voids. Our city's gardening life is part of a much larger narrative, and I want to breathe life into that.

Gardening is a very personal, intimate activity, and gardening stories can be precious treasures that can only be heard with a degree of trust. I've been entrusted with some wonderful stories of this ilk that I think give this book a fresh angle. It is probably more about gardeners than about gardens.

Introduction

Avant-garde filmmaker and gardener Derek Jarman once said, 'Paradise haunts gardens, and some gardens are paradise.'[1] That paradise haunts gardens is an inescapable fact for garden historians, not merely because the word 'paradise' derives from the ancient Persian word for a garden or enclosed space. When people garden, they create their own paradise; they organise nature in particular ways; they carve out space from whatever lies beyond the enclosure they have made. This is true even when the garden is but a few pots of geraniums or herbs on a terrace.

It is true also for colonies formed on islands. Islands, especially those in the Pacific, were frequently mythologised as 'paradise' within the European imagination.[2] Aotearoa New Zealand, which emerged for Europeans out of Terra Australis Incognita, an imagined great continent that turned out not to exist, was conceptualised early on as a paradise, a potential garden. In the 1840s, the era of systematic colonisation, New Zealand was marketed in Europe as a kind of Arcadia, an unpeopled wilderness of tremendous natural abundance ripe for tapping by hard-working men: the perfect site for a perfect garden.[3]

But, of course, Aotearoa was not unpeopled, and the inhabitants had been cultivating gardens (ngā māra) for centuries. The earliest Polynesian settlers carried precious seeds, tubers and cuttings from their homeland in order to establish food gardens as soon as they made landfall. In diverse and ingenious ways they developed a strong and sustainable culture of food production, evidence of which has survived the 200 years of land transformation that has resulted from European colonisation: the proliferation of towns in which most households have at least a small section, and farming practices that have drastically altered the majority of the countryside.

In New Zealand garden histories, however, the topic of Māori gardening is largely glossed over at best. Although the value that Māori food production added to the new European colony in the early 1800s has been well documented, there has been a collective 'writing out' of Māori from garden histories from that point on, as if at this juncture Māori simply stopped gardening. Despite the tremendous challenges of the nineteenth and twentieth centuries, however, some Māori gardening rituals and practices survive.

While many older New Zealanders will remember home gardening as the norm, for many younger citizens the age of having a colourful front garden facing the street and a large back yard crammed with fruit trees, vegetable beds and a chicken run is an almost mythological era. This concept of *rus in urbe* – 'country in the town' – came under pressure with the fashion for patios and swimming pools, and later from subdivision projects and the advent of townhouses and apartment-style homes with little or no garden space. As a result, many New Zealand suburbs are becoming relatively garden-free zones; even in subdivisions where new homes are surrounded by expanses of highly fertilised lawn, sites of intensive food production are virtually non-existent.

A garden can tell us much about the residents of a property. New Zealand gardens of the early 1970s, wrote Austin Mitchell in *The Half-gallon Quarter-acre Pavlova Paradise*, were like immaculate market gardens, each with 'a crop large enough to feed the entire Vietcong for decades'.[4] They were immaculate because 'most New Zealanders spend Sunday afternoons driving round the suburbs looking at everyone else's homes and gardens'.[5] Janet Frame described suburban Auckland gardens of the late 1970s as having 'shaven lawns ... with no grass blade out of place', and 'carefully mown grass verges', swimming pools, pebble gardens, gnomes and 'trees that burst in summer with delicate ... blossoms'.[6] Keri Hulme described a West Coast garden: 'A neat lawn bordered by concrete paths. No flowers, no shrubs. The places where a garden had been were filled with pink gravel.'[7] Although poetic, fictional creations, these examples suggest just how tightly a resident's identity may be bound to their garden.

Gardening is as subject to fashion swings as it is to changes in technology or economic needs, but these fashions tend to apply most meaningfully to the gardens of the wealthy. Many garden histories over-emphasise the importance of 'trends' in gardening, perhaps because these are a simple way of showing how gardening has changed over time. Keith Sinclair has suggested that the generic 'quarter-acre' section with its house and garden has been a crucial site for the development of the New Zealand identity;[8] and the notion of a generic garden attached to a generic 'Kiwi house' has been identified (for example for the period 1920–60 by James Belich[9]). But there are widely divergent approaches and meanings attached to these gardens.

Three elements are lacking from the writing on New Zealand home gardens. The first is that few localised studies of garden-making in New Zealand towns and cities have been attempted.[10] Specific, local studies would allow us to better understand the regional similarities and differences of gardening practices around the country.

Second, there has been little recognition of the way gardening practices sometimes signify resistance to a dominant culture. Where public scrutiny demands, for example, flower gardens of a particular type, other forms of gardening may reveal radicalism. Where beautification of townscapes had become a male concern by the end of the nineteenth century, the campaign by women for more vegetable gardening added a gendered twist to the situation. Michael King noted the 'radical' nature of the Wellington Vegetable Club of the 1940s.[11] Clubs like this could be home to discussions not only about *how* but also *why* to grow vegetables; often this was about achieving independence from the prevailing economic paradigm.

Third, little has been written about the ecological effects of garden-making.[12] The aggregated efforts of thousands of home gardeners must have a significant environmental impact. Gardening practices that are water-intensive, fossil-fuel hungry and poison-based are simply not sustainable. Widespread recognition of the need to protect the environment is vital, and began to emerge in New Zealand around the turn of the twenty-first century.

In this book I begin an examination of these three elements in the garden history of Aotearoa. By exploring the often-dramatic difference between what garden guides were suggesting and what ordinary gardeners were actually doing, I offer an implicit critique of a certain method of doing gardening history – that of relying on the garden guides to explain the ways people gardened. More important for many gardeners is the strength of family and other social networks that support the retention and transmission of gardening knowledge and experience. As these networks weakened during the second half of the twentieth century, the impact on home gardening was profound.

The influence of wealthy garden patrons in our communities has been immense, but I do not focus on the oft-documented gardens of the well-to-do. Nor is this a study of commercial gardening – and for that reason I touch only lightly on the significant contribution of Chinese market gardeners. I do not examine public gardens, or the gardens of asylums, cemeteries and the like. Instead, I explore the historical processes behind gardening modes which, consciously or otherwise, often work against those forms articulated by horticultural societies and the popular press, who were so often drawn from a cast of white males representing the loci of political power. I explore the ways in which cultural meanings have been inscribed in the land through gardening practices.

In this book I examine what have variously been called 'ordinary', 'domestic', 'home', 'labourer's', 'working class', or even 'vernacular' gardens.[13] None of these terms quite encapsulates what is meant here, in part due to periodisation and cross-cultural issues. Margaret Willes encountered a similar problem in her 2014 book *The Gardens of the British Working Class*; in an effort to broaden her scope to pre-Industrial Revolution gardening she invoked the contemporaneous term 'lower orders' (meaning 'the poor').[14] Such labels are even more problematic when writing about the gardens of a colonised people, not least because of the concept, alien to Māori, of individually owned parcels of land. I examine the ways that the gardens of labourers, the working class and poorer members of society – people affected by forms of exclusion from the exercise of political and economic power – sometimes mitigate the effects of that exclusion.[15]

Perhaps 'non-elite gardens' is the best description. My preferred term is 'common' gardens, because this not only incorporates the idea of 'ordinary' or 'vernacular', but also hints at the notion of 'the commons' and the communalism that can underpin the gardens of those with the fewest material resources. In this connection, 'common' may speak equally to traditional Māori approaches to gardening and to late twentieth-century urban community gardens.

Agropolis: urban food production emerging from earthquake rubble in central Christchurch, 2014.

Photo: Eroica Ritchie

I also explore the development of a 'green cities' movement in New Zealand, and in particular the growth of community gardens and the Transition Towns phenomenon that surfaced in late 2007. In the wake of fruit and vegetable price increases of more than 12 per cent in the 12 months to the end of 2008, garden centres noted sharp increases in the sale of vegetable seeds, seedlings and fruit trees. This seemed to signal a renaissance in home gardening. A similar trend in new community garden projects, often of a small scale, suggests that the New Zealander's love of gardens has revived and is returning, albeit in new forms. We are seeing the development of a new kind of mahinga kai – food-gathering places – in the wastelands of concrete and tarseal that are our cities.

I have given priority to those gardening activities that help articulate the differing social and cultural circumstances that have prevailed over time. I have examined the economy of resources – including ideas and knowledge – that have impacted the garden forms that consequently emerged. And I have sought to give voice to those gardeners who have so often remained voiceless.[16]

CHAPTER 1

New food in the gardens: 1800–50

Gardening has a long-established history in Te Ika a Māui and Te Wai Pounamu, the two main islands of Aotearoa New Zealand. Evidence of the roughly 600 years of Māori gardening that preceded the arrival of Europeans may be found in many parts of the country. In the Taranaki area, for example, the names of the mountains Pouākai ('pillar or source of food') and Kaitake ('source or abundance of food able to be grown on the slopes') connect them with ancient gardens.

Tradition tells us that the ancestors of Māori navigated their way to New Zealand in large sea-going waka or canoes, from a place called Hawaiki in East Polynesia.[1] One of these was the *Aotea*, captained by the chief Turi. After establishing shelter at Pātea, in Taranaki, Turi set about developing food gardens with his people, planting the kūmara and other foods that they had carried across the ocean.[2] Local tradition suggests that the earliest ancestors of these iwi, Te Kāhui Maunga, wrote their gardening into the landscape as genetic stock from this first garden was distributed throughout the area.[3]

Historians and archaeologists have divided pre-European Māori gardening into the initial colonisation or early garden period (AD1200–1500) and the late garden period (AD1500–1800).[4] It is impossible to say where the first gardens were established, but Louise Furey's thorough analysis of identified sites indicates that the greatest concentration of Māori gardens occurred in Northland and Auckland. These appear to date from

1400; garden sites at Palliser Bay, on the south coast of the North Island, date from 1350. Evidence of gardening in the cooler South Island is less common, and little has been found south of Banks Peninsula.[5]

Over time, early Māori adapted their gardening traditions 'to accommodate local circumstances and environmental conditions' in New Zealand.[6] Archaeological evidence from Ahuahu/Great Mercury Island, off the northeast coast of northern New Zealand, suggests cultivation may have commenced there using a perennial or low-intensity system, which changed to an annual or high-intensity system from 1500. The plants the first settlers brought with them were mostly marginal, even in the sub-tropical regions of Aotearoa; on Ahuahu the absence of taro pollen and the presence of kūmara starch granules from 1500 may denote an increased reliance on the more climatically suited kūmara.[7] It is possible that Ahuahu functioned 'as a nursery or experimental garden before the expansion of crop production to other areas across New Zealand'.[8] Evidence suggests the expansion of dryland kūmara cultivation on the mainland came later, after 1600.

Tāmaki-makaurau

A good place to begin is Tāmaki-makaurau (Auckland), the name of which alludes to the desirable resources of this once richly fertile volcanic isthmus. Early settlement centred around the volcanic cones, where Māori established extensive cultivations of some 2000 hectares or more.[9] Radiocarbon dating of material from Matukutūruru, a cone also known as Wiri Mountain, points to a first occupation there around AD1200–1300, although Furey considers this too early. The nearby Puhinui garden appears to date from no earlier than 1600.[10] Archaeological evidence suggests gardening at Wiri was a key activity from the outset.[11] Substantial middens, indicators of early habitation, show that many people lived amid these gardens.[12]

The area immediately surrounding the isthmus was also worked. Ōtuataua, on Manukau Harbour, was intensively cultivated from about the fourteenth century.[13] Of the original 8000 hectares of worked land, the remaining 100 hectares (now the Ōtuataua Stonefields Historic Reserve) reveals a complex system of mounds, walls and terraces that demonstrate some of the ways by which Māori transformed their environment to create microclimates for crops.

Remains of garden stone-wall complex at Ōtuataua, Auckland.
Author photo

Slightly to the north, in the suburb of East Tāmaki, archaeological exploration has revealed the remains of a garden that was maintained for a short period in the mid-sixteenth century. Within the complex was an oven, indicating that gardening and cooking occurred in the same space for this community. There is no evidence of walls or fences, which suggests defence was perhaps not an issue.[14] Extensive terraced gardens in nearby Mt Wellington also appear to have been undefended. Mt Eden to the north was similarly covered in settlements and gardens.[15]

The most famous of Tāmaki's kūmara gardens was at Maunga-kiekie (One Tree Hill), a volcanic cone that was entirely converted into terraces surrounded by extensive earthworks.[16] The lowest – and unfortified – areas of these were worked gardens.

These small snapshots show some of the key features of Māori gardening in the thirteenth to eighteenth centuries. Early Māori employed diverse methods to get the most out of the sites they cultivated. They undertook significant environmental modification to maximise production, and were skilled at preserving crops in storage pits to ensure a supply of seed for the following season. The sheer scale of their gardens indicates the huge labour investment required of these communities.

Kūmara pits at the top of Three Kings, 1899.

Stephenson Percy Smith, Boscawen Album, Auckland Museum, PH-NEG-B5451-C10438

While there is evidence that some areas were occupied only briefly, in general iwi appear to have remained within the isthmus for a long time. The lack of evidence of defences – particularly in the later period – suggests that despite the large numbers of tribes inhabiting the area, peace prevailed. This should be no great surprise: peace is certainly a prerequisite for a strong gardening culture to develop as it did in Tāmaki-makaurau. In the 1740s, however, this relative peace came to an end with the incursion of Ngāti Whātua and their displacement of Te Waiohua and those who maintained alliances with them. By the end of the eighteenth century the population of the area had shrunk by up to one third.

Landscape modification

Gardening communities such as this, albeit on a smaller scale, existed in many parts of Aotearoa, at least as far south as Banks Peninsula. Archaeological records reveal a variety of cultivation techniques. Furey, following Walton, summarises these as stone structures (including rows, alignments, mounds, heaps and stone-faced terraces), ditches and channels, borrow pits (from which sand or gravel has been removed for horticultural purposes) and modified soils.[17]

Remnants of garden walls at Flea Bay, Banks Peninsula.
Author photo

From her archaeological study at Palliser Bay, Helen Leach determined that 'gardening was being practised … from the time of first settlement at least to the end of the sixteenth century'.[18] Examination revealed different types of gardens, including flatland or sloping beds – both of which were divided into strips by trenches or stone rows, sometimes two or three metres tall – and terraced gardens.[19] Stone walls have been discovered throughout the North Island and as far south as Banks Peninsula; those present at Palliser Bay would have aided plant growth by radiating heat and creating wind breaks, and suggest the value these growers placed on orderliness.

The ditched swamp garden, of which the Far North provides good examples, is another style of Māori environmental modification. Evidence of these is visible from the air and through examination of soil depth and composition, as Susan Bulmer has described.[20] Rows of raised beds were separated by controlled water channels; water-loving plants like taro could be grown in the wettest areas and other crops on the higher levels. Rotted vegetable material cleared from the drains was placed on top of the beds to ensure ongoing soil fertility. Around 1500 hectares of these swamp gardens were developed around Kaitāia as early as the seventeenth century, probably making use of existing natural water courses. Another, at Motutangi to the north, covered about 50 hectares; here the sloping part of the garden was irrigated with the aid of a spring, the water guided as necessary by blocking off certain ditches.[21]

The practice of altering soil composition to suit particular crops was widespread: 33 te reo Māori names for different soils have been identified in the Auckland region alone.[22] Gardeners with access to seaweed would have applied it directly to the soil as a fertiliser, and gravel, sand, charcoal and other materials were sometimes mixed with soil for drainage or fertility, or to better retain solar warmth. Leach suggests soil modification was common practice in Māori gardening from at least the thirteenth through to the nineteenth century.[23] At Pohatu pā (now Flea Bay) on Banks Peninsula, part of the Koukourarata rohe or district, flecks of charcoal mixed with small rounded stones and shingle from the beach can be found in the soil up to about 75 metres above sea level over an area of around 15 hectares. The name of the pā can be translated as 'the place of stones', and the archaeological evidence evokes an image of a hillside of kūmara plants growing in fields of small stones.[24] In the Bay of Islands, Hare Te Heihei's grandfather Tareha learned about growing kūmara in gravel from someone who 'always got better crops by bringing gravel up from the river … They all cultivated kumaras in this manner and found the yield greatly increased.'[25]

Soils modified with gravel and charcoal at Pohatu pā, now Flea Bay, Banks Peninsula.
Author photo

Evidence in the landscape also reveals other forms of environmental modification, most notably forest clearance by fire. Burning cleared ground for planting and, as Prebble and others have discussed, enriched soils with charcoal. It could also expose soils on hillslopes, which were then 'driven and held in small catchments for rapid garden construction'.[26]

Crops

Polynesian settlers brought breadfruit, bananas and coconuts, but these plants were not suitable for the Aotearoa climate. They also brought taro, kūmara, gourds and uwhi, a kind of yam. Taro was the staple food source in tropical Polynesia in 1200, and Prebble suggests 'there was a strong cultural impetus to maintain taro as a staple crop' for the earliest settlers in Aotearoa, too.[27] From around 1600, however, dryland kūmara production seems to have expanded, which may reflect a gradual acceptance of taro's incompatibility with the climate. Although better suited to the cooler temperatures, especially the subtropical climate in the northernmost parts of the country, even kūmara was a marginal crop in many parts. Its popularity is reflected in the fact that 45 varieties were recorded in 1888.

Māori also cultivated a number of native plants, including the cabbage tree, tī kōuka, a valued food plant that was treated with reverence. Ngāi Tūāhuriri rangatira Wiremu Te Uki made this clear: 'If any one of us even a Māori set fire to any of these cabbage trees, he would be killed at once. That is our law.'[28] Many stories exist on the subject of tī. Terry Ryan believed cabbage trees were planted as signposts for mahinga kai (food-gathering places) throughout the swamplands of pre-European Christchurch, although this is difficult to verify.[29] In various parts of Canterbury tī was cultivated in plantations: Te Maire Tau has explained how the settlement of Tuahiwi, just north of Christchurch, was known for its abundance of cabbage tree cultivations.[30] Māori produced several cabbage tree cultivars for food, and these were sometimes planted in groves to attract birds.[31] Other food plants grown in Māori gardens included the fruit-bearing trees karaka and poroporo, and para (a fern) and the rock lily rengarenga, the roots of which were eaten.[32]

Food was a priority – of course it was – but some fibre plants and decorative plants were also tended. Three examples are well documented. Elsdon Best observed that Māori planted New Zealand flax or harakeke, which yielded fibre for making clothing, baskets, nets and many other items, 'adjacent to their villages'.[33] Varieties of flax suitable for different purposes were grown apart from one another.[34] The paper mulberry (aute), introduced by Māori from Polynesia, was cultivated to make bark cloth, but was rare by the time of Cook's first visit.[35]

There is something of a scientific consensus that the beautiful flowering 'kākā beak', now all but extinct in the wild, has survived only because it was cultivated in Māori gardens. The first specimens were collected by Joseph Banks and Solander in 1769, when it was already a rarity in the wild. But kākā beak was widely grown around Māori villages.[36] Although it is a nitrogen-fixing legume, there is no evidence to suggest that it was grown for this role. One story suggests it was propagated to feed caged tūī, which then attracted other birds – a useful food source.[37] Another suggests it was grown for its tasty seedpods.[38] John Clemens notes an 1855 reference to kākā beak flowers worn as decoration.[39] Sensitive to frost, the kākā beak requires particular care and stands out among New Zealand flora as unusually colourful: it is quite possible that it was grown simply for its beauty. The difficulty writers have in understanding why this plant was cultivated by Māori may be an example of the prejudice Māori gardening has experienced.

Tools, pests and rituals

Frequent tilling of the soil was a key method of keeping down weed pests, and hard woods from the forest, such as maire, mānuka, matai and akeake,

A kō, a wooden implement for digging.
Museum of New Zealand Te Papa Tongarewa, MA_I029204

A woman uses a timo, c. 1900.
Museum of New Zealand Te Papa Tongarewa, MA_I088065

Taumata atua – a stone resting place for Rongo-marae-roa.
Museum of New Zealand Te Papa Tongarewa, MA_I029479

A whakapakoko atua or 'godstick'.
Auckland War Memorial Museum, 1929.32

provided timber for a number of garden implements.[40] Probably the best known of these is the kō or digging stick, principally used for breaking up soils for planting. A takahi or teka, a footrest, was often lashed to the stick with harakeke or other fibre to enable the user to apply force. Men working with the kō sang particular chants, partly to bring rhythm to their activity.[41] Other tools – wooden kāheru (spades), timo (grubbers) and hoto (shovels) – would have appeared familiar to European eyes.

Early European visitors noted that cultivation was carried out by 'the commune'. Elsdon Best describes how all members of a community 'took part in some of the tasks of clearing and preparing the ground; planting, tending and gathering the crop. Chiefs, warriors, commoners, slaves, women, old and young, all assisted in some way.'[42] These were truly 'community gardens'.

Growing healthy crops also meant successfully participating in the greater cosmological order. Mauri – stone images intended as the resting places of gods – and whakapakoko atua ('godsticks') were common objects in gardens, and intricate rituals, largely the domain of rangatira and tohunga (chiefs and priests), existed around aspects of the horticultural cycle.[43] The most important were to do with kūmara. The deity Rongo-marae-roa was responsible for agriculture, and Māori regarded the kūmara as Rongo-marae-roa's child, and therefore sacred.[44] In 1820 Richard Cruise observed that, at planting time, temporary huts were built within the kūmara grounds; no one was allowed to 'pass the boundary night or day until their labours [were] terminated'.[45] Ernst Dieffenbach described this vegetable as 'the food most esteemed among the living'; he observed that women engaged in kūmara cultivation were considered tapu and 'must pray, together with the priests, for the success of the harvest'.[46] The first planting (māra tautāne) was always set aside for the gods, the atua, and accompanied, as Best recorded, with the 'recital of certain *karakia* ... by a presiding priestly adept ... to attain the will of the gods'.[47] Planting was marked with a high degree of ritual; the men wore particular clothing, and a tohunga chanted a karakia while placing the tubers on prepared hillocks.[48]

Māori gardeners faced problems from a number of insects and birds. Of the former the most difficult was āwheto, the sphinx or convolvulus hawk moth. In its caterpillar stage, for which 19 te reo names exist, it feeds voraciously on kūmara foliage and can devastate crops.[49] Women

Pātaka (storehouse) at Te Parapara, Hamilton Gardens.
Author photo

were often tasked with picking caterpillars off the plants, as were tamed and tethered seagulls. Another method of eradication involved smoking the bugs out by burning green leaves, such as kawakawa.[50] In Taranaki various methods of reducing pest numbers have been recorded, such as using smoke, crop rotation and leaving ground fallow.[51] In the Waikato area, stories tell of keeping invasive bird pests such as pūkeko at bay by rattling shells on a string, a job given to the elderly.[52]

Religious rituals also existed for keeping pests at bay. Samuel Marsden recounted a story of a plague of caterpillars that attacked a kūmara crop. '[The] priest came and stopped several months; he performed his religious rites, and directed every principal cultivator to make an ark for his god.' Each 'ark' was 'about five foot long by two foot wide, and eleven inches deep' (150 x 60 x 28cm). It was 'painted and ornamented with carving and various figures', and in it were placed 'sacred provisions'. In this case, the caterpillars left soon after.[53]

Māori closely observed the celestial bodies, something Best described extensively in his 1922 book *The Astronomical Knowledge of the Māori: Genuine and empirical*. Tūhoe chief Tūtakangāhau provided Best with a glimpse of one ceremony in which young, new growth was gathered by priestly adepts and 'offered to the stars'. Certain ritual language was intoned and the stars were asked to provide abundant harvests and to prevent any damage to the crops.[54] The Pleiades cluster, known to Māori as Matariki, gave strong cues of what to expect for the coming season when it appeared at the beginning of the gardening year; and various other planets, stars and the moon were also believed to influence gardening behaviour.

One cannot generalise about the relationship between gardens and Māori life for the whole of New Zealand, as there were significant regional variations. None is starker than the fact that kūmara was rarely grown south of Banks Peninsula because the climate was too cold. Canon Stack's reminiscences paint a vivid picture of the care taken to nurture this important crop:

> To regulate the temperature of the soil, and to secure perfect drainage, they covered the surface of the *kumara* plantations with fine gravel to a depth of six inches, which was afterwards formed into mounds about two feet in diameter, and arranged over the field with the precision of the squares on a chess board, and in these mounds the *kumara* tubers were planted. Breakwinds of *manuka* branches, varying from two to four feet [60–120cm] in height, were erected every few yards apart, and in such a way as to secure the largest amount of sunshine and shelter to each plant.[55]

Taumutu, just south of Banks Peninsula, was perhaps the southernmost kūmara-growing area, and although a few references suggest that Māori did manage to grow the tuber further south in places with amenable microclimates (for example Timaru and Wānaka), there was certainly nothing on the scale seen in the north.[56] For most southern hapū, gardening was just not a major activity; their emphasis was on hunting and gathering. Fish, birds and various plant foods – roots and berries mostly – figured prominently in their diet. Some of these were useful items for trading with gardeners further north.[57]

Where gardening was carried out, however, it was central to Māori life – and the records of European visitors from the late eighteenth century bear this out. During his first voyage to New Zealand in 1769–70, James Cook observed 'plantations of yams, sweet potatoes [kūmara], and cocoas [taro]' in northern parts of the country.[58] Of the first plantation the voyagers explored, just north of Tolaga Bay, Joseph Banks wrote:

I have seldom seen land better broken up. In them were planted sweet potatoes, cocos [taro], and a plant of the cucumber kind, as we judged from the seed leaves which just appeared above ground. The first of these were planted in small hills, some in rows, others in quincunx, all laid most regularly in line … The cucumbers [gourds] were set in small hollows or ditches, much as in England. These plantations varied in size from one to ten acres each. In the bay there might be 150 or 200 acres [60–80ha] in cultivation … Each distinct patch was fenced in, generally with reeds placed close by one another, so that a mouse could scarcely creep through.[59]

European explorers recorded similar plantations up and down the North Island, and noted that while the soil was good and the plantations extensive, crop diversity was limited to kūmara, gourds and taro.

The introduction of European vegetables

The first Europeans to visit Aotearoa, observing that Māori were accomplished gardeners, gifted them with seeds and tubers brought from Europe. French explorer Jean François Marie de Surville supposedly left the first potatoes, in 1769. Jules Crozet, who was on this voyage, wrote that he himself had 'formed a garden on Motuaro Island in which I sowed the seed of all sorts of vegetables, stones and the pips of our fruits, wheat, millet, maize … everything succeeded admirably'.[60] Cook and his navigator Tobias Furneaux established several gardens in Queen Charlotte Sound in 1773, in which they planted potatoes and several types of corn, beans, kidney beans and peas.[61] These horticultural experiments had long-lasting consequences. While visiting the sounds in 1839 Edward Jerningham Wakefield discovered 'wild carrots and turnips which remain as relics of our great navigator'.[62] Likewise Rev Henry Hanson Turton, visiting in late 1846, found an area 'covered with deep grass, carrots, parsley, & cabbage, doubtless the remote production of Cook's horticultural labours'.[63] These were most likely from Furneaux's garden: in 1777, while lamenting the loss of his own plot, which he said had been removed in his absence, Cook pointed to the success of Furneaux's, saying, 'we found cabbages, onions, leeks, purslain, radishes, mustard, etc., and a few potatoes'.[64] Despite receiving no attention from Māori in the area, the vegetables had survived.

By the turn of the century potatoes had become a valuable crop, which Māori supplied to hungry crews on visiting ships in exchange for iron items such as axes.[65] Captain Wilson reported extensive fields of potatoes at the Firth of Thames in 1801, and others noted 'immense' quantities of potatoes growing in the Bay of Islands by 1805.[66]

New crops were fast becoming incorporated into the traditional māra, or gardens. Early in 1815 Samuel Marsden described 'one field which appeared ... to contain forty acres [16ha], all fenced in with rails and upright stakes tied to them to keep out the pigs. Much of it was planted with turnips, common and sweet potatoes, which were high in cultivation.'[67] But the new food plants had not entirely displaced the old. Marsden noted that kūmara, with all the sacral qualities ascribed to them (not to mention, we have to suppose, their flavour), were still 'esteemed by [Māori] as the choicest food' in 1814, a point he reiterated in 1819.[68]

In the less hospitable climate of the far south, where the cultivation of food had been relatively rare, a process of adaptation was also under way that would usher in an era of unusual prosperity for Māori. Southern communities apparently became proficient gardeners in a short space of time. In 1809 one European observer in the Foveaux Strait area recorded that the diet of local iwi consisted principally of fish supplemented by potatoes; in 1813, a 40-hectare potato field was recorded near Bluff, and other gardens of potatoes and cabbages were recorded nearby in 1823.[69] By 1828 potatoes were 'widely available' in the area.[70]

Potato harvests were also transforming the Māori way of life in Otago.[71] According to Atholl Anderson, by the 1830s the major Māori economic activity 'seems to have been the cultivation of potatoes for shipment to Sydney, and consignments of 25–60 tons were sent regularly during the 1830s and 1840s'.[72] Around Banks Peninsula Māori traded potatoes as early as 1821 and were still trading them in Akaroa in the late 1830s; and in the Queen Charlotte Sound area in 1839 Edward Jerningham Wakefield exchanged pipes for potatoes.[73]

A key change in gardening at this time involved the declining value of kūmara in the Māori economy. A temperamental crop requiring, as we have seen, windbreaks, careful soil preparation, pest control and occasionally protection (palisades to keep out either marauding neighbours or introduced pigs were recorded by Edward Shortland in 1842 at

Kirikiriroa Pā), as well as a great deal of ceremony, the kūmara could not compete with the more hardy and prolific potato.[74] The potato revolution encompassed the whole country, and by the 1850s the transition from kūmara to potato as the main crop was complete.[75]

Māori were increasingly growing many other crops as well. As early as 1815, Samuel Marsden's travelling companion John Nicholas recorded a village where adjacent gardens were stocked with kūmara, potatoes, gourds, cabbages, turnips, maize and taro.[76] In 1820 near Waiau Bay, Richard Cruise observed a group of people standing around a two-month-old pea plant, which they had fenced and declared tapu, 'and the greatest care appeared to be taken of it'. The seed had come from the ship *Coromandel*, which was in the area collecting kauri.[77] He noted that cabbages were among the foods commonly eaten. By the end of his visit, late in 1820, many chiefs were growing peas that they promised to save for seed, watermelons 'were in great luxuriance', and there was even an orange seedling that was 'rigidly tabooed by the inhabitants'.[78]

In 1827 Augustus Earle recorded 'plantations of Indian corn, Kumara … and potatoes' at Taiāmai in the Bay of Islands, while at nearby Kawakawa in the same year, French explorer Dumont d'Urville observed carefully planted 'kumara, potatoes, maize, melons, and pumpkins'. By the mid-1830s Tūhoe hapū in the Bay of Plenty had acquired 'wheat, corn and potatoes' and 'other European fruit and vegetables', and at Puketī Pā Te Makarini and his people were growing 'large cultivations of potatoes, maize, and kumara'.[79] In 1839 there Ernst Dieffenbach found potatoes, taro, pumpkins, onions and watermelons growing in a 'clean and well-weeded garden' in the 'middle of the bush', and noticed tobacco growing in one cultivation.[80] And in 1840, two days before Te Tiriti o Waitangi was signed, New Zealand's first surveyor-general Felton Mathew came across a large population of Māori in the Bay of Islands subsisting almost entirely on cockles supplemented with 'potatoes and cabbage, both of which grow in profusion, as well as turnips and radishes, whether indigenous or derived originally from the European settlers I cannot discover'.[81] At the foot of 'Pouki-nui' near 'Waimati', Mathew visited a pā where the people had cultivated a large area 'exceedingly well' in corn and potatoes.[82] At a 'native camp' in the Hauraki Gulf he was offered melons, 'calapashes' (gourds), onions and potatoes.[83] John Williams recorded Māori cultivations of kūmara, potatoes and corn in 1842 around the Bay of Islands, and at Tauranga he noted potatoes and corn 'in great abundance'.[84] That year Charles Terry added pumpkins, maize and watermelons to the list of introduced crops being harvested.[85]

The Māori people's attention to these crops drew early admiration from European visitors. Earle wrote: 'To view the cultivated parts from an eminence, a person might easily imagine himself in a civilised land.' The thought that the land and people actually *were* civilised was of course ridiculous to Earle, but he couldn't help but observe that:

> for miles around … nothing but beautiful green fields present themselves to the eye. The exact rows in which [Māori] plant their Indian corn would do credit to a first-rate English farmer, and the way in which they prepare the soil is admirable.[86]

Samuel Marsden described a Māori kāinga or village in North Cape with 'pretty little cottages and their gardens high in cultivation, neatly fenced and laid out, and the potatoes, yams, etc., all planted in separate beds with not a weed to be seen'.[87] Describing one Māori potato garden at Ōkuratope in 1815, he observed that he 'had never seen finer potatoes under the best culture'.[88] John Nicholas, who was travelling with Marsden, was similarly impressed: 'the extreme regularity that was observable in the whole process did much credit to their taste in agriculture. The plants were all disposed in perfect order, and the weeds rooted out with minute exactness.'[89]

Frederick Spencer also described Māori gardens: 'The Maori's staple was the Irish potato, but many other articles entered into the diet, such as Taro,' he wrote. He recorded that kūmara at the time was 'almost always a long thin tuber which when cooked was of a jelly consistency and sweet. Its cultivation was very intricate it was shielded by rows of manuka branches about four feet [1.2m] apart and the patches were kept well weeded.' Besides noting the skill with which Māori tended their crops, he observed changing practices: 'The calabash gourd … was … eaten [and] furnished the principle vessel for carrying water in or cut longitudinally it served as a plate or bowl. It has practically disappeared. Then Pumpkins and vegetable marrows were largely grown and later on maize was grown and later still wheat …'[90]

These gardens were a godsend to arriving Europeans, and were sometimes the target of European theft. One whaling crew entered the

'*Whare and garden*'. *Rows of corn beside Māori dwellings, c. 1900–20.*
Auckland Libraries Heritage Collections, 1572–1407

gardens of the chief Taraia near Waimate in the Bay of Islands around 1815 to steal potatoes. Incidents such as this could end in bloodshed.[91]

Potatoes produced more food per hectare than kūmara, and these extra provisions allowed large war parties (taua) to be fed in ways that were unimaginable before. As Cruise noted in his reflections on a visit to New Zealand in 1820, the food needed by traders, missionaries and their attendants could 'only be obtained from the natives, who, in a certain degree, are able to command the trade they choose to take'. The result of this, he saw, was that 'the most murderous weapons [i.e. muskets] have been disseminated among them'.[92] The humble potato became a critical factor in the 'arms race' that preceded a series of tribal battles between 1818 and 1840, known as the Musket Wars.

There were environmental consequences to this developing economy, too. In 1820 the captain of the whaling ship *Saracen* gave muskets to Māori in Whangaroa as he sailed out; in exchange the Māori would plant more potatoes to feed his men upon his return. As a result, Cruise recorded, 'the whole country was this day in a blaze' as the locals set fire to the land – a precursor to cultivation – on a scale hitherto unseen.[93]

Diversification of crops also meant gardening communities now existed in colder, more southern locales than ever before. Initially, these

Māori attempted to incorporate potato cropping into their cycle of seasonal movement.[94] But by the 1840s, as Anderson has noted, many Māori had 'become drawn into sedentary market gardening' as a result of European commercial demands. This would have implications for the entire Māori economy.[95]

A changing economy

Europeans arriving in Aotearoa did not always comprehend that Māori had a long gardening history of their own, and some even attributed Māori gardening prowess to their own influence. In a letter dated 1841 Rev H.H. Turton, writing from the Aotea Mission Station, reported that the local Māori had fed 'upwards of a thousand visitors with food'. The food consisted of pork, potatoes, kūmara, Indian corn and fish, 'with a few indigenous fruits'. He deemed this impressive for a people who, in his opinion, 'were but lately the most inhuman of Cannibals & at perpetual war amongst themselves'.[96] In Turton's mind the civilising influence of European culture had created the prerequisites for the successful cultivation of land.

Some new arrivals, unaware of the extent of adaptation that had taken place in a short space of time, remarked on how quickly they felt Māori had managed to pick up gardening skills in this new era. In 1844, as Harry Evison has pointed out, European visitors commented favourably on the evidence of Māori agriculture from Foveaux Strait in the south to Banks Peninsula, where by 1848 the cultivations at Puari extended to about 101ha.[97] Visitors to one farm on the peninsula brought back 'magnificent "cobs" of Indian corn, perfectly developed ... and some water-melons also perfectly ripe, both of which the Maories [sic] had grown in their gardens'.[98] Peaches were popular. Mary Hobhouse wrote that the English settlement in Motueka in 1859 was sparsely populated; the only European trees in evidence were peaches and weeping willows, 'the former greatly preponderating'. 'Maoris have found out the merit of peaches & their pah here rises from a thicket of brilliant pink, or I would say is buried in it.'[99]

New settlers throughout Aotearoa New Zealand relied on local iwi for many things, including food, and trade was strong. Sarah Mathew recalled Māori selling potatoes and kūmara in Auckland in the early 1840s, and in 1853 'potatoes, wheat, maize, melons, grapes, pumpkins, onions, flax' were available, along with some livestock and firewood.[100]

A trick photo depicting Hoane Parāone Tūnuiārangi as both a gentleman and a gardener, holding his produce. Date unknown.

When famine struck, as it did in 1850, the Māori people supplied potatoes to Europeans in need.[101]

By 1850 when the planned settlement of Christchurch was 'planted', Māori were well versed in trade with Europeans. As hungry colonists arrived, Māori obliged. Some settlers, like Edward Ward, struggled to do business, however, as cultural misunderstandings got in the way of transactions. In his diary on 1 January 1851 he commented, 'Bought ... a mat of new potatoes from a Maori. These folk are very hard to deal with, as they ask exorbitant prices, and don't understand being beaten down in English, so that one is forced to walk away in despair ... They have pigs, peas, potatoes and poultry for sale, and plenty of buyers if they would only ask reasonable prices.'[102]

A productive garden meant economic independence and a ready source of food, and new settlers established these as quickly as possible. Many sought Māori assistance with the task. At Marsden's mission station at Kerikeri in 1819, Māori were engaged to sow maize and a 'number of seeds', which had been brought from England via Sydney, and they planted grapevines and fruit trees.[103] In 1823 Marianne Williams, wife of missionary Henry Williams, noted on arrival in Paihia near Waitangi that a large garden had already been established for them, with 'numerous rows of peas and beans'. In 1844, she employed 'native boys' to work in the garden, including 'Takapui', who helped to plant potatoes: 'Natives at work in the garden,' she wrote in her journal; 'Native gardeners busy cutting and trimming the grass border.'[104]

As they developed their own spaces of self-sufficiency, Pākehā became increasingly frustrated with Māori trading arrangements. In 1852, for example, rangatira George Williams Metehau said he desired 'that white men should dwell at my place at Te Tuahiwi [north of Christchurch]. But it depends on the payment; £8 a-year is the payment for dwelling in the land, to cultivate potatoes, wheat, corn, and all other seeds.'[105] The new colonists did not appear keen on such proposals. Metehau's offer was ignored as colonists set about draining their new properties – situated as these were among 'inconvenient' wetlands – and grew their own food instead, something Māori were possibly not anticipating.

Sentiments expressed by European settlers in the 1840s and 1850s present a changed perspective from the laudatory remarks of eighteenth-century explorers and earlier missionaries. Pākehā began to view Māori people as somewhat careless and lacking in gardening skills. Charles Heaphy, on board the *Tory* when it arrived at Port Nicholson in 1839, perceived their cultivation methods to be 'very defective', and complained that when growing potatoes Māori 'never plant the "eye" separately, but reserve the very smallest of the produce for the next year's seed, and not unfrequently depend on what remains in the ground, after a negligent search, for the ensuing year's crop'.[106] Some began to voice more extreme views about Māori laziness. John Cracroft Wilson was dismayed that Māori around Christchurch 'reside for the most part on their reserves & they work these for themselves sufficiently to sustain life'. In his opinion they were, 'with a rare exception here & there a lazy race, & are in my opinion inferior in energy even to an Asiatic'.[107] Opinions such as these allowed Europeans to justify their presence and position in relation to Māori.

Māori inhabitants there were already cultivating vegetables in significant quantities in Port Nicholson, later called Wellington, when it was selected as the site of the first of the New Zealand Company settlements in 1839. In some places Europeans simply established themselves over Māori gardens. Government buildings butted onto significant Māori cultivations at Pipitea pā, which were later swallowed up in the burgeoning suburb of Thorndon.[108] Archaeological evidence suggests there may have been a garden on the banks of the Avon River in what is now central Christchurch: the area where a stone gardening implement was discovered was the site of the first land office, established by the colonists from 1850; the provincial government buildings were later built nearby.[109]

Māori approaches to gardening had changed dramatically by the mid-nineteenth century, but their world-view had not; their spiritual relationship with gardening was steadfast and their long-standing horticultural knowledge remained strong. One family's story exemplifies this.

Hēnare Tomoana, of Ngāti Kahungunu descent, was a Hawke's Bay rangatira born in the 1820s or early 1830s. Although enthusiastic about

Māori man sorting kūmara.

Ref: 1/1-006227-G. Alexander Turnbull Library, Wellington, New Zealand. / records/22313821

land sales and a government supporter during the campaign against Te Kooti in 1868, he later changed his position. As a politician (1879–84) he argued for the implementation of Te Tiriti o Waitangi and the abolition of the land laws.[110] He himself had experienced the dispossession of his ancestral lands around Waipatu and Wairua near present-day Hastings.

Hēnare was a skilled gardener. His son Paraire Tomoana, born in 1875, inherited both his father's land situation and his gardening knowledge. He retained traditional knowledge of how to cultivate kūmara, and an understanding of gardening according to lunar cycles as taught by his parents. He held the deep belief that gardeners were part of the wholeness of creation; working as part of a greater scheme, they were giving something to the land, and could never own it.[111] He later published a Māori lunar gardening calendar for planting and harvesting.[112] He and his wife Kuini Rīpeka Ryland in turn instilled this knowledge and values in their own children. A prevailing sense of wholeness remained intact, in this one family at least.

Storing kūmara and potatoes.

Weekly News. (1910) East Coast. Storing kumara & potatoes. *Auckland War Memorial Museum, neg. C10787*

Just as many of the mid-nineteenth-century colonists considered Māori gardening – by then confined to reserves at the edges of the new settlements – as somehow irrelevant, so too have many garden historians turned their focus from Māori gardens to European gardens from that point, without looking back. In their view, Māori gardening has largely been seen as a sideline to the British main act. In 1971 Robert Cooper claimed the 'story of early Auckland gardens begins in the journal of Sarah Louisa Mathew', who wrote about her bulbs in 1840.[113] This view has been presented even in relatively recent work.[114] Such assertions are rooted in the marginalisation of Māori.

The Musket Wars and subsequent alienation of Māori from ancestral gardening lands throughout Aotearoa in the second half of the nineteenth century had a significant impact on the ancient gardening culture of Aotearoa. When Te Tiriti o Waitangi was signed in 1840, notwithstanding the impacts of the Musket Wars, the future for Māori gardening did not look so bleak. Māori had been extremely adaptive over the previous 50 years, just as they had after their arrival some 600 years earlier. They had incorporated many new foods into their gardens and had developed techniques to grow them well. Their gardens had supported the establishment of new power bases, which, ironically, now included the settlements of incoming Pākehā.

CHAPTER 2

The fruitful paradise: 1840–1914

Potatoes were well established in Māori gardens by 1810. Māori readily adopted new edible plants and by the late 1830s, when European settlers began arriving to establish a new life in Aotearoa, iwi were cultivating a number of familiar vegetables. In the decades that followed, British settlers added a selection of fruit trees, projecting their hopes for a rediscovered Eden onto New Zealand's accommodating soils. A few flowers were included in settler gardens to remind the new colonists of home, but these were likely secondary to the establishment of the vegetable and fruit crops necessary for sustenance. The specific forms of these colonial gardens had much to do with class – the size of the garden, which plants could be procured and from where – but the functions of gardens were similar across settler society. In the making of these gardens, colonial women found new ways to negotiate gender relations as well. Māori, on the other hand, would find themselves increasingly confined to reserves on the fringe of new settlements, their long-established relationships with mahinga kai unexpectedly brought to an end. With this came a loss of mana.

Introducing variety

The colonisation of Aotearoa by British subjects was facilitated by the New Zealand Company, which did its best to promote the fecundity of the new land and the possibilities the country offered of social mobility in a new society. The New Zealand Company was established to realise the visions of its progenitor, Edward Gibbon Wakefield, who expounded a theory of planned settlement that revolved around the idea of a 'sufficient price' for land. In Wakefield's plan, the price of land would be such that the wealthy would be able to afford it outright; those who could not would be able to work for them until they could also afford to purchase land in the not-too-distant future – thereby establishing a functioning economy.

Complicating this process was, of course, the matter of land ownership. Sensing that the British government was moving to annex New Zealand, in 1839 the New Zealand Company commenced a process of pre-emptive land acquisition from Māori. After all, they needed land to fulfil the promises they had made – of sufficient land for all – to the settlers who were planning to emigrate. In 1840, with the signing of the Treaty of Waitangi, the annexing was complete and the British government tightened the process of land sales; it would later contest the New Zealand Company's purchases. Nevertheless, and despite considerable financial trouble, the New Zealand Company established settlements in Wellington and Wanganui (1840), Nelson and New Plymouth (1841), Dunedin (1847/48) and Christchurch (1850). In doing so, they transplanted a particular kind of class structure, a set of social dynamics, and aspirations of freedom and independence that at the time were utterly out of reach to many in Britain.

The new colonists set about planting vegetables and fruit trees from home. More varieties were flowing in all the time, largely from Australia. By 1842 Wellington gardeners could purchase the seeds of peas, dwarf French beans, cabbage, broccoli, celery, parsley, kale and turnips, and a range of fruit trees – pears, apples, plums, cherries, peaches, nectarines, figs and apricots – direct from Sydney.[1] By the 1850s Wellington and Nelson had become important sources of plant stock – especially fruit trees – for the other provinces.[2] William Tye, who arrived in New Zealand in his mid-twenties in 1842, quickly established a garden in Auckland in which he sowed turnip seed, planted peach stones and raised melons, beans and cucumbers. Just once he noted in his diary that he had set flower seeds: like many others, his chief focus was on food production.[3]

There was a great deal of boosterism in public discourse regarding the efforts of the first settler gardeners. Charles Heaphy, who was employed as a surveyor, topographical artist and draughtsman for the New Zealand Company, was impressed with productivity: 'Several specimens of giant cabbages have been produced from land near Wellington, and the size of one grown in Robinson Bay, close to the beach of the harbour, would have, in England, obtained it the premium of an horticultural society,' he crowed. 'To prove the extraordinary productiveness of the soil even in places where it should be least expected ... I saw in the garden of a poor man on the beach, strawberries, with ripe fruit, growing in the sand, within ten paces of the sea.'[4] Fruit trees held great promise and were prolific: 'From a few peach and apricot stones that had been planted in the northern parts of the island, there are now great numbers of those trees growing wild about the Hokianga and Bay of Islands.'[5] In the Wellington area there were already 'many fruit-trees in blossom, most of which had been obtained from Sydney'.[6] Heaphy described the numerous varieties growing further north:

> In the garden of a timber merchant residing on the Hokianga,
> I noticed 106 different species of vines, which included all the most
> celebrated in Europe; and they all appear to be flourishing. In the
> same garden were all the English fruits, – pears, cherries, plums,
> gooseberries, both 'Cape' and common, currants, raspberries, and
> mulberries, growing in the utmost profusion.[7]

Heaphy's observations are in line with other contemporary descriptions of gardening in New Zealand. In 1844 John Williams remarked that in the Bay of Islands, 'all tropical and temperate vegetation thrive remarkably. Figs and vines in fact everything which can facilitate or afford gratification to the human family ... [the] exuberance of the geranium is beyond description; perhaps not more so than the rose.'[8] Later he remarked on the proliferation of apples, pears, peaches, plums, cherries, apricots, figs, grapes, strawberries and cape gooseberries: 'every kind of fruit propagates luxuriantly'.[9] Heaphy, Williams and others were reifying the propaganda used earlier by the New Zealand Company as part of its sales pitch to new colonists.[10]

Rev Henry Hanson Turton described his Aotea Mission Station garden in 1842 with similar emphasis: he had '[s]trawberries, raspberries, mulberries, currantberries, English & Cape Gooseberries – vines, figs, pumpkins, cucumbers, water & musk melons – Trees – apple, pear, cherry, peach, quince, plum & apricot'. He had raised many of these from stones and seeds, including almonds, which he bought from the grocer. He added a weeping willow, 'shrubs & flowers in abundance', and '[p]eas, parsley, beans, celery, onions, turnips, cabbages, cauliflower, radishes, sages, thyme, Hobart town potatoes &c'. Turton had established his garden by removing the existing fern root, and maintained it with the help of a Māori man called Piripi, who sowed and transplanted the vegetables.[11] A plan of these mission grounds shows the extent of land set aside for potatoes and other vegetables, and their proximity to the 'native cultivations'.

Pūtaringamotu

In the area that became Christchurch, this pattern was mirrored closely. At Pūtaringamotu (probably meaning 'a place to catch forest fowl', though it is often translated as 'place of the severed ear' or 'place of an echo'), the local hapū of Ngāi Tūāhuriri leased 'the land running six miles [9.6km] in every direction' to the Deans brothers, John and William, from 1846.[12] This had at one time been a notable Waitaha settlement.[13] For Ngāi Tahu it was a settlement and mahinga kai, with a 'proper fort'.[14]

The brothers transformed this rich site and named it Riccarton. In January 1844 John Deans wrote to his father to say that about three-quarters of an acre (3000m²) had been cleared and planted with 'cabbages, peas, potatoes, onions, leeks and parsnips'. His carrots, turnips, melons and cucumbers had unfortunately been 'eaten up by a small fly'. The brothers had established many fruit trees and strawberry plants. 'Next year,' John stated confidently, 'we expect to have plenty of strawberries and perhaps a few apples and plums.'[15]

The abundance of the garden at Pūtaringamotu was a key strand in this correspondence: 'Our garden crops were all very good last season,' John wrote in 1845. 'We had about twenty very good apples on one tree and one plum which proved to be a greengage. In a year or two we should have plenty of apples, plums, cherries and peaches.' The strawberries were producing, although not abundantly, and they had planted gooseberries.[16] In 1847 John reported that the garden continued to be productive: 'We have this year more than a dozen apple trees loaded with fruit, a good many plum, cherry and peach trees, all with more or less fruit, and a great many young ones coming in.' Recent additions included more gooseberries, currants, pears 'and a few roots of rhubarb'.[17]

Two years later the brothers began propagating fruit trees to sell to the wave of settlers who were expected to arrive in 1850.[18] Their garden earned 'the admiration of all the new colonists on account of the luxuriance in growth of everything in it'. The trading of trees, bushes and currant canes for 'valuable seeds and trees' was a central interest in 1851, and in February 1853 John wrote that the 'garden looks uncommonly well. The apples, plums and peaches are abundant.'[19] In March, 'seldom a day goes over that we do not carry a bushel or so' of fruit; and in October the fruit trees were 'the admiration of everyone'.[20] In 1854 he recorded an 'abundant crop of gooseberries, currants'.[21] Riccarton, the first European garden on the plains, was a great success.

John Robert Godley, founder of the Canterbury settlement, was impressed by what he saw at Riccarton. The garden, he wrote to his father in 1850, 'which never saw or heard of manure, is producing luxuriantly every kind of vegetables and fruits. I never saw a finer show of them.'[22] John's wife Charlotte, who was concerned with beautification around Christchurch, involved herself with the garden at Riccarton. Her primary stimulus for establishing a garden of her own was the expense and availability of fresh vegetables: 'There is no certain supply of them here, and the new potatoes are still 2d. a lb.'[23] In a letter to her mother in 1851 she commented on some recently arrived recipe books, noting that the recipes 'must be a little modified here, where a cabbage, for instance, costs a good deal more than a pound of the best beef or mutton; sixpence or even ninepence'.[24] Economic autonomy was vital, and a garden was one means of achieving this.

The wind made cultivation problematic in Canterbury. Charlotte noted that 'things grow so fast under a fence, or any protection against the wind'.[25] Forest remnants were an instructive feature for early gardeners, something Cracroft Wilson alluded to in 1854. While discussing the need for shelter in Canterbury, he drew attention to the example set by the Deans brothers at Pūtaringamotu, whose garden 'lies to the North East of the Riccarton Bush one of the few bits of Forest which has escaped the grass fires of Yore. The fruit trees in it are forwarder than those in any garden in Christ Church, & doubtless could they speak, they would say, a South Wester was not such a terrible thing after all.'[26]

Colonial independence

Self-sufficiency was of paramount importance to the first settlers, and they rapidly transformed 'waste land' into gardens – the 1852 census recorded 24 hectares of fenced-in colonial gardens within Christchurch alone.[27] One writer in 1900 looked back at the 1850s: '[A]t the risk of being egotistical it may be asserted that all of the Pilgrim Fathers, the leaders and the rank and file alike, were of the true British stuff, filled

A Māori man tends a vegetable garden, c. 1910–20.
Photo George Bourne, Auckland Museum, George Bourne Album 328p2n1

A European woman in a piupiu and tiki stands in bare feet in a vegetable garden, c. 1910–20.

Photo George Bourne, Auckland Museum, George Bourne Album 328p2n2

with indomitable pluck and energy.' They were determined 'to persevere until they had made the desert wastes of the Canterbury Plains blossom as the rose'.[28] In the south, some felt that Ngāi Tahu, who had stepped up their production of European foods in an attempt to participate in the commerce of the new colony, charged too much for their produce.[29] The convenient myth of superior colonial productiveness enabled the negation of Ngāi Tahu's gardening efforts and, by 1900, in the popular imagination, Māori gardens might never have existed. Through the creation of self-sufficient settler spaces, Māori were effectively excluded from economic activity.

Local organisations everywhere pressed settlers to establish food gardens as quickly as possible. In more than one 1851 editorial, the *Lyttelton Times* urged everyone to participate: '[L]et every poor man hire his acre or two of land, and cultivate it during his spare hours.'[30] The writer suggested a reward be offered for effort: 'The man who, next autumn, will shew the largest quantity of human food for the capital which he has expended since his arrival, ought to receive the honour of a civic crown from his fellow colonists.'[31] The newspaper's column 'Errors of Immigrants' warned of the futility of growing crops that were not immediately useful, and called for maximum productivity: 'We are anxious, and we think reasonably so, to see every settler a grower to a greater or lesser extent. Every kitchen garden, every poor man's acre, in course of tillage we hail as an additional reason to hope for prosperity as a settlement. Everything depends on the extent to which we are producers.'[32]

The extraordinary gardening diary of William Trotter of Lower Hutt, near Wellington, shows how seriously this message was taken on

The vegetable garden at a cottage in Waiuku, 1868, with cabbages, onions, currant bushes and a fruit tree.

Auckland Libraries Heritage Collections, Footprints 07077

board. Between 1850 and 1861 he worked in his garden on almost every day except Sundays. His main activities centred on food production. In September 1851 Trotter spent his time planting potatoes, making a melon bed, mounding potatoes, digging fruit borders, cutting wood, training and grafting trees, tying and digging raspberries, 'dressing' (oiling) apple trees, and sowing cucumbers, radishes, cauliflowers and tobacco.[33] Aside from a few scant references to his flower garden, the only concession to something other than a form of economically useful gardening was his regrettably unexplained 'Labyrinth', which received a fair share of his attention for a number of years before disappearing from his records without explanation.

Trotter occasionally recorded births, deaths, marriages, earthquakes and major weather events in his diary, and accompanied most of these short entries with heartfelt prayers. After a violent storm in 1855 he wrote, 'O God thou all powerful Being make me remember this day until the day of my death.'[34] Describing a major flood that was 'within 6 inches of coming into our house' and which had 'done a great deal of injury to many of the Settlers', he asked that God help him 'to submit with humility to all his wise dispensations still believing that all things work together for good for his chosen people'.[35]

Trotter's self-transcendent observations remind us of the hardships of establishing a new life in a strange land, and help moderate the view that colonisation was primarily about conquest and domination. Indeed, Trotter's 'Edenic' view – in which he prioritised caring for creation – was shared by many European settlers in this period, as Beattie and Stenhouse have shown.[36] Recreating Britain by gardening was not Trotter's prime motivating factor; rather, his prayerful reflections on natural disasters suggest his personal trajectory was more spiritual than imperial. For Trotter, at least, the unsurpassed fecundity of his garden, conjoined with his pronounced surrender to his God, conjures up more of a commitment to recreating Eden.[37]

Ōtautahi

The ancient Māori site of Ōtautahi, in Christchurch, is an equally important example of a site where Canterbury colonists learned to establish crops in the new conditions. The rangatira Tautahi had visited the area regularly from his home at Koukourarata (now also known as Port Levy) on Banks Peninsula, and his people gathered birds there after the long journey through the swamp. Tautahi himself was married at Ōtautahi and was later buried nearby, giving his name to the location and, subsequently, to the whole area of Christchurch. The site was included in the 5.5 million hectares that Ngāi Tahu sold to the Crown in 1848 in a settlement known as Kemp's Purchase. Under the terms of the purchase, the iwi understood that such food-gathering places would be set aside for their use.[38]

Ōtautahi was a critical site for European occupation in Canterbury as it was at the uppermost navigable point of the river Ōtākaro, which settlers renamed the Avon River. Here Edward Jollie completed the survey of Christchurch in November 1849 and spent his evenings fishing for eels, pig-hunting and shooting birds.[39] On his 1850 'Plot of Christchurch' he marked out a proposed 'botanical garden': a site of 9.3 hectares situated in a bend of the Avon. According to W.G. Brittan, whose farm was nearby, the area was at that time covered in tutu trees, fern and grasses.[40]

The 'botanical garden' was cleared and maintained as Christchurch's first plant nursery by William 'Cabbage' Wilson, and quickly became the main site on which European plant materials were propagated for distribution.[41] So important was the enterprise to the success of the settlement, in Godley's eyes at least, that Wilson was initially allowed to live there rent-free.[42] Wilson did not disappoint Godley, and in his advertisements in the *Lyttelton Times* he offered a wide range of available food plants. His first advertisement on 13 September 1851 listed seeds of 'Blue Scimitar and Bishop's early Dwarf Peas, Green Windsor Beans, Globe Onion, Carrots, Parsnips, Turnips, Celery, Parsley, Asparagus'. Trees for fencing, shelter and fruit –'Ribstone Pippin Apple, Kentish Cherry, Green Gage Plum, and Brown Turkey Fig' – were also available.[43] The following month he advertised for sale 1000 fruit trees, 900 red and white currant bushes, roots of asparagus and rhubarb, and seeds of onion, peas and beans.[44]

In early December Wilson listed:

5,000 Early YORK CABBAGE. 5,000 Early Sugar Loaf do [ditto].
100 Red Pickling do ... 3,000 Green Curled Savoys, 2,000 Cauliflower, 2,000 Early White Cape Brocoli, 2,000 Grange's White do.
1,000 Cabbage and Cos Lettuce, 1,000 Green Curled Endive.[45]

Wilson recorded that 13 acres (5ha) of his garden were planted in cabbages, and his skill with the brassica soon earned him the nickname 'Cabbage Wilson'.[46]

Ōtautahi had been entirely transformed, but Māori still regarded it as an important food-gathering site. In 1868 – coincidentally the same year that Wilson became mayor – Ngāi Tūāhuriri rangatira Hākopa te Ata o Tū lodged a claim with the Native Land Court in which he reminded people that as a mahinga kai, Ōtautahi was not included in Kemp's Purchase.[47] His claim was dismissed, however, because the Crown had already sold the land.[48] Wilson's garden at Ōtautahi was a palimpsest: the resources of the site remained constant but their form, and those who had access to them, changed dramatically.[49]

Edible imperatives

In 1852 the *Christchurch Guardian* and *Canterbury Advertiser* ran a series of gardening and farming columns by Wilson, 'specially prepared for the *Guardian*', which the *Lyttelton Times* continued later that year. Wilson's major thrust was the promotion of economic gardening in the fertile though windy environment. His columns provide a useful record of the crops known to settlers. On 3 June 1852, in early winter, Wilson advised:

> The only seeds which can safely be sown are Onions, Radishes, Lettuces, Mustard, and Cress; and within the shelter of a bank or close paling, having a Northern aspect, a few Early Frame or Early Charlton Peas, and Early Mazagan Beans may be sown ... Plants of Early Yorks, sown late in Autumn, may now be transplanted ... Plants of Cauliflowers, sown at the same date, should now be fit for transplanting ... Asparagus and Rhubarb roots may be planted ... The sets of Potato Onions may now be planted in beds, 3½ feet wide, with four rows in each bed, and nine inches from set to set.[50]

Wilson urged readers to plant fruit trees with short stems and 'hoop trained' for added hardiness in difficult environments.[51] Gooseberries, apples, pears, plums, cherries, peaches, nectarines, figs, currants, raspberries and strawberries all received mention.[52] In August he suggested sowing dwarf pea varieties due to 'the high winds of midsummer and the scarcity of Pea-stakes', and in September he recommended the cultivation of onions, carrots, parsnips, turnips, cabbages, cauliflowers, Brussels sprouts, 'Curled Kail, or Borecole', spinach, curled cress, mustard, radishes, lettuces, peas and beans, beetroot, flowering broccoli and celery.[53] Asparagus, sea kale, horseradish and globe and Jerusalem artichokes were listed, and a number of herbs: parsley, summer savory, sweet marjoram, sweet basil, bush basil, marigolds, thyme, sage, winter savory, pot marjoram, balm, borage, spearmint, peppermint, hyssop, lavender, and 'such Medicinal herbs as Rosemary, Feverfew, Penny-royal, Chamomile, and Horehound'.[54] His lists included a small number of decorative plants: mignonette, sweet peas, balsams, nasturtiums, 'Convolvulus Major', roses, fuchsias, hollyhocks, dahlias, chrysanthemums, pinks and carnations, 'and the many other varieties of Biennial, and Perennial Herbaceous flowering plants, of which the settlement, young though it is, already contains a very creditable collection'.[55]

The personal records of gardeners of the 1850s show that the initial drive towards food production was maintained. Charles Bridge established Opawha Farm at the Māori site of Ōpawaho in 1850 and gave a good, although brief, account of his gardening work for 1852, all of which revolved around food crops.[56] William Guise Brittan's garden, not far from Wilson's nursery, was sheltered by hawthorn and 'furze' plants and supported 'an abundance of vegetables and fruit trees of many kinds'. His vegetables in 1852 had all 'succeeded to perfection'.[57] To the east down the Avon River, Henry Tancred had a 'small kitchen and flower garden, with a few fruit trees', while further downstream, Percival had half a hectare of cultivated garden 'full of vegetables of all kinds, and fruit trees'.[58] Two labourers on the properties of Westenra and Wilkinson each had a small potato garden.[59] Conway Rose noted in 1852 that he had not yet laid out his garden in central Christchurch, but that his vegetable crops were thriving: 'We have plenty of potatoes, turnips, cabbages and beetroot growing round the house.'[60]

Robert Bateman Paul in his 1857 *Letters from Canterbury* provided an overview of the success of these early gardeners. Following the establishment of fast-growing shelter trees, fruit and flowers were doing well, he wrote. Mr Barker had produced 200 'fine peaches from standard trees', while Mr Thompson had excelled in growing 'some bunches of out-of-door grapes'. In his estimation, the success of the settlement was due to the work of its gardeners.[61]

Not everyone was impressed, however. John Cracroft Wilson was frustrated by the effort immigrants were putting into making gardens at the expense of doing a decent day's work for landowners such as himself. In 1854 he complained that 'no labourer in Canterbury thinks of coming to his work before 8 o'clock am or remaining at it after 4 o'clock pm'. According to Wilson, a working man might get up in summer at 4am, 'work hard for three hours in his own garden or field', then turn up to work 'listlessly enough, for his Employers for 8 hours' at a relatively high rate, before returning home to give 'his own garden or field the benefit of two additional hours' good hard labour'.[62] This was largely academic – Wilson himself had a number of Indian servants to do this work for him – but it is worth noting that the dismissive attitudes that both Hall and Cracroft Wilson had earlier directed at Māori were now also being directed at the labouring classes.

Edward Gibbon Wakefield, the founder of systematic colonisation, eulogised the gardens of Christchurch when he visited the new settlement in 1853. To his wife Catherine he wrote that the 'neighbourhood is beauty itself', and in a letter to Henrietta Rintoul he remarked on the 'beauty and natural fertility' of Canterbury.[63] To his friend Robert Rintoul he praised the food crops: 'vegetables at Canterbury were finer than I have ever seen before'.[64]

Horticultural societies and civic duty

Horticultural societies were quickly established in the colony – in Wellington by 1842, in Auckland by 1843, and in Dunedin and Christchurch in 1851; and horticultural societies and officialdom – whether the New Zealand Company, the Canterbury Association or the government – were tightly connected.[65] Godley's interest in housing Wilson rent-free on public land despite the payment schedule laid out in the lease agreement is evidence of a close relationship between the Canterbury Association and the horticultural society (variously called the Horticultural and Agricultural Society and the Botanical Society). At a meeting of the Botanical Society in 1852, at which it was proposed that the society be renamed the Christchurch Agricultural, Botanical and Horticultural Society, Godley was elected president. Brittan, who ran the Canterbury Association's Land Office, became treasurer.[66] The Canterbury Association as manifested in Christchurch effectively *was* the horticultural society. Edward Jerningham Wakefield, Edward Gibbon's son and former secretary, had switched from the Wellington Horticultural and Botanical Society and by 1852 was on the management committee of the Canterbury Botanical Society.[67]

Horticultural society shows were important sites for the construction of civic identity, since the production of plentiful vegetables and fruit equated to being a good citizen. The awarding of prizes for food production was a way of recognising the efforts of colonists in developing their gardens for the greater good. The Wellington Horticultural Society show in 1842 offered prizes for a range of vegetables, fruit and flowers.[68] The first produce exhibition in Christchurch, in 1852, had categories for fruit, flowers (including native shrubs and flowers) and vegetables, the latter incorporating 'potherbs', native grasses and comb honey; a second competition in the autumn focused exclusively on food.[69] This pattern continued throughout the 1850s and into the 1860s in the burgeoning Canterbury settlement. The Auckland Horticultural Society ran a similar contest in 1861 with 11 categories for pot plants, seven for flowers, 36 for fruits and 28 for vegetables.[70] The following year most categories were doubled in size.[71]

William Wilson used his *Guardian* column to promote the Christchurch society's produce exhibition on 16 December 1852:

> The return of August – the first month of Spring, and the prospect of a Horticultural Exhibition, are mutually suggestive of the numerous Gardening duties which the present month imposes; for they who would secure the superior advantages afforded by early cropping, as well as they who would endeavour to maintain the fertile character of the Canterbury Plain – by exhibiting its choicest vegetable productions at the forthcoming exhibition – must each consider that they have no time to lose; that *immediate* planting and sowing are essential to secure in high perfection most of our Fruits, Flowers and Vegetables, by the 16th December – the anniversary of our Settlement.[72]

The fact that a produce exhibition was held to mark the second anniversary of the arrival at Lyttelton of the first four ships that brought settlers to the area suggests that gardening and civic duty were enmeshed even at this early date in public discourse.

Beautification agendas: 1860s–1914

By 1865, however, the emphasis of these exhibitions had shifted. Although abundant food gardens were celebrated, the horticultural society's desire to promote beauty was becoming more prominent, and dovetailed with the expansion of global plant exchange networks, as James Beattie has shown. Many flowers of Asian and especially Chinese origin were welcomed into certain New Zealand gardens at this time.[73] This internationalism probably reflected technological advances such as the development of steamships and of Wardian cases, a type of sealed protective container for live plants, as well as the expansion of imperial power through the nineteenth century which, as Nigel Rigby has argued, allowed 'far greater freedom of access to botanical specimens'.[74] 'Cottage garden' flowers – lilliums, hollyhocks, verbenas, geraniums and 'marygolds', all common in English gardens, but originating in Asia, North America and Europe – were predominant at the Christchurch show of 1865.[75] Fruit was still well represented, naturally, and was considered by judges to have been 'particularly good': Mr Potts exhibited a basket of 32 varieties of fruit, a phenomenal number by today's standards; Mr Lance presented some 'very fine tomatoes'; and Mr McCormick won a prize for two bunches of grapes from his hothouse.[76] In Auckland the pattern was similar. In 1872, 36 categories were offered for pot plants and 27 for flowers, while fruit and vegetable categories had shrunk to 12 and 15 respectively.[77]

Over time the exhibitions gradually increased their promotion of flowers. There were exceptions of course: in 1876 and 1877, for example, the Auckland society organised a judging of working men's vegetable gardens, but this attracted only 11 entries and was discontinued.[78] Horticultural shows were eventually referred to simply as 'flower shows', and the Canterbury Horticultural Society was squeezed out of existence by the rose and chrysanthemum societies in the 1890s.

A detailed analysis of property advertisements for the Christchurch area (see overleaf) shows where the gardening bias of ordinary home-owners lay. From a total of 2338 properties listed in the *Press* between 1865 and 1965, roughly half yielded specific information about gardens. Apart from 1865, when shelter trees were well represented, fruit was the most frequently listed garden element until just after World War One. Where gardens were mentioned in the period 1895–1915, orchards and

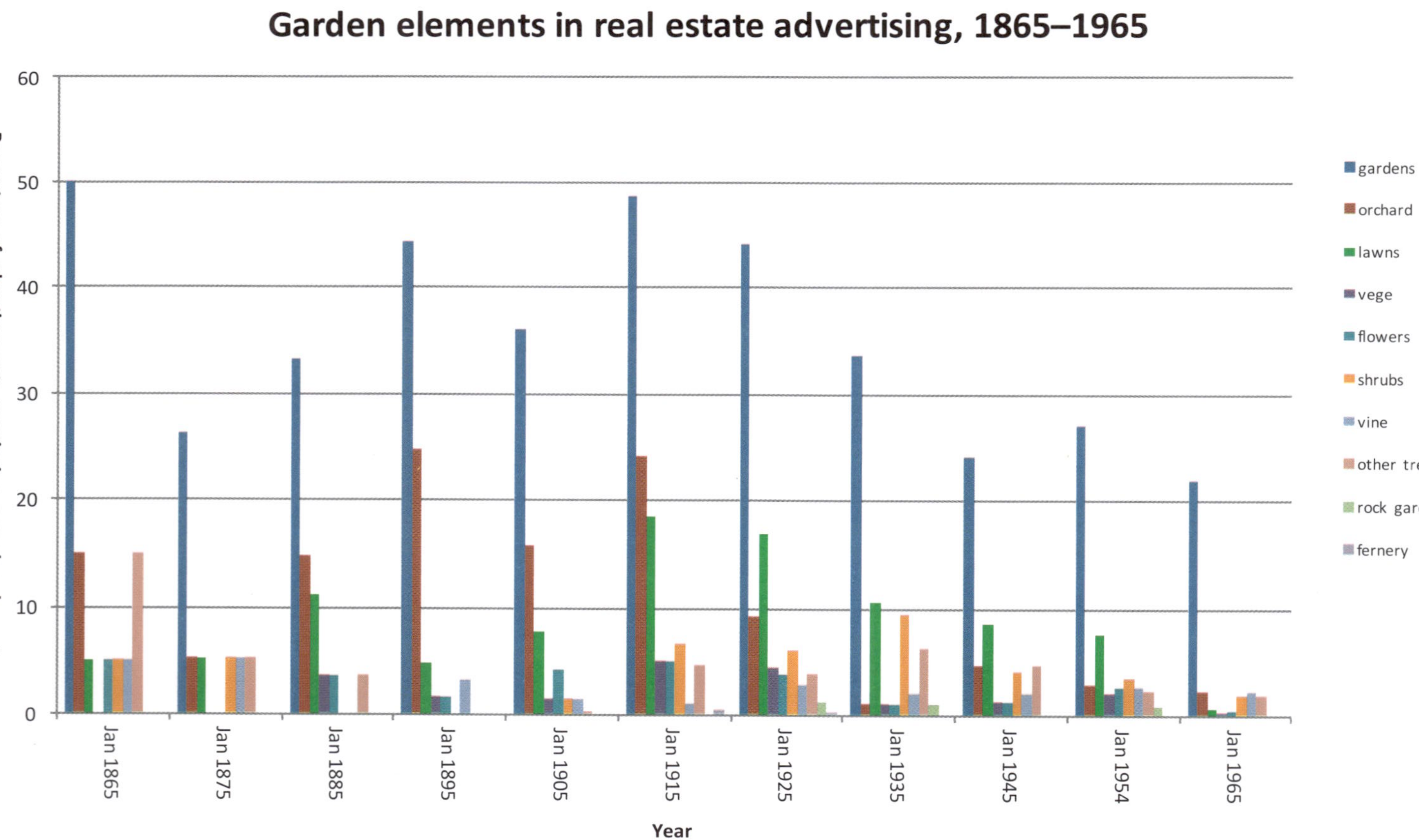

Garden elements in real estate advertising, 1865–1965
Percentage of advertisements mentioning a garden element
60
50
40
30
20
10
0
Jan 1865
Jan 1875
Jan 1885
Jan 1895
Jan 1905
Jan 1915
Jan 1925
Jan 1935
Jan 1945
Jan 1954
Jan 1965
Year
gardens
orchard
lawns
vege
flowers
shrubs
vine
other trees
rock garden
fernery

*A boy holds a cat under a fruit tree at the house of Charles Dawes in Umawera,
c. 1890–1910.*
Auckland Libraries Heritage Collections, 1142-D105

fruit trees were the most popular descriptors. Home food production
was clearly still a primary concern for Christchurch residents, at least
until 1915.[79] The 'abundance ethic' seems to have been an active force,
even if the horticultural establishment no longer officially endorsed it.

This pattern was not confined to Christchurch. Alexander Stuart
bought 54 hectares of bush-covered land on Otago Peninsula in 1863
and certainly busied himself with food production in preference to
beautification. His diary over the following few years lists the ongoing
sowing and planting of various vegetables common to us today, as well
as unusual ones like giant asparagus and sea kale.[80] His interest in food
production extended to food preparation also, as his journal shows:
included in it are recipes for mushroom ketchup, preserved cucumbers
and rhubarb wine. A rare gardening catalogue from 1866 produced by
Thomas White of Lambton Quay, Wellington, likewise suggests that
although the emphasis of horticultural societies was changing, the
demands of the average gardener were not: it listed 123 varieties of vegetables and a mere 70 flower types.[81]

Gardeners' records through the 1870s and 1880s continue to reinforce this emphasis. The diary of J. Stanley Monck, who settled in the Redcliffs area (Te Rae Kura) in Christchurch at what became Moncks Bay, provides ample details of his garden activities from 1869 and 1875. His primary emphasis was on edible crops. In July 1869 he planted early potatoes, sowed grass seed, turnips and peas, made a new strawberry bed, purchased rhubarb roots and cabbage plants from 'town' and undertook a lot of digging.[82] Later in the year he planted pumpkins and peas, harvested strawberries and cherries and dug potatoes for Christmas.[83] In 1871 Monck began beautifying his garden, constructing flowerbeds in April and October of that year.[84] In November he noted 'mowing grass in garden' and 'drove to town bought [sic] back rose trees'.[85] He planted blue gums in 1872, constructed paths round the house and sowed Cape broom seed 'round wife's garden' in 1873, fenced the garden in 1874, and spent part of June 1875 'marking out paths etc' in a new area.[86] In the meantime, however, he maintained a considerable degree of self-sufficiency in food. Fruit trees, grapevines and strawberries continued to receive large amounts of attention, while potatoes, rhubarb and currants occupied ever-increasing space.[87]

The abundance ethic characterised New Zealand gardens and is demonstrated in another diary, that of a Hawke's Bay station gardener. In July 1873 the writer received trees and shrubs from Napier, planted 15 drills of 'potato onions' (a multiplying onion) and 10 rows of cabbages (some with stable manure), anointed the apple trees with kerosene and pruned the peaches. Trees of economic value were being established at the same time as the orchard: the writer took delivery of plums, cherries, pears, apples, filberts, ashes, oaks and a mixture of pines.[88] In 1876 he received 37 varieties of fruit trees from Auckland.[89] On one day in November 1874 he sowed numerous flower seeds, including sweet william, German asters, marigolds, poppies, sweet alyssum and a nasturtium, but this nod towards decoration is hardly characteristic of the garden he was working on, where the overwhelming emphasis was on vegetables and trees.[90]

A similar list of fruit trees exists in an anonymous scrapbook from 1883. Although its place of origin is a mystery, inserted newspaper cuttings on gardening from the Auckland area suggest it may have been compiled there. An orchard plan within – 'Plan of orchard in front of

William McCullough in the newly planted orchard at the McCullough homestead on the corner of Lauderdale and Eskdale roads, Birkdale, 1896.

Auckland Libraries Heritage Collections, B0421

house Planted 1883' – lists 16 trees (figs, quinces, apples, plums and pears) and a grapevine.[91] Two pages of the scrapbook depict 44 varieties of fruit trees set out in rows.[92] Whether this was a merely a pipe dream is unknown, but the fact that such a list was compiled in itself tells us something about the aspirations of this particular gardener.

This scrapbooker was not the only one with large plans. In Waihola township near Dunedin, a three-acre (1.2ha) property advertised for sale in 1896 boasted 100 fruit trees, including apples, pears, plums and cherries, and 100 gooseberry and currant bushes.[93] Nor must we assume that such impressive orchards could be found only in the hinterlands; in Christchurch in 1905 one Sydenham garden was described as having 'profitable fruit trees' and 'a lot of raspberry bushes, etc', while a quarter-acre (0.1ha) property in Richmond housed 80 fruit trees.[94] It is interesting to note that the Onehunga Horticultural Society's autumn show of 1890 had an excellent selection of apples, quinces and lemons, but the flowers were poor. The excuse given was that the weather had been bad, which might have been plausible except for the fact that the

'Tom Baker's plum trees', c. 1900–20.
Auckland Libraries Heritage Collections, 1572-1451

greenhouse plant section attracted no competitors at all. The garden bias of the good people of Onehunga could not have been clearer.[95] Thus, the desire to grow food first is reflected in the gardens of both the wealthy and the humble Pākehā populace. And by this point, there were some very grand estate gardens around the country.

There was a gendered element to this work. Annette Bainbridge has shown that in early settler society, at least in Canterbury, women were significant contributors to the gardening efforts in middle- and upper-class homes: 'Gardens were for many women places of freedom, action and resistance, and the centre of family life and their social networks.'[96] Bainbridge points to the garden essentially as a space of performing and negotiating gender identities. At the same time, she notes how important the kitchen garden was for women, who bore the main responsibilities of providing food for the family.[97]

Samuel Daws and Ann Peet Daws with baskets of apples and grapes, 1903.
Auckland Libraries Heritage Collections, 1572-1051

As the examination of property advertisements for Christchurch and records of gardeners in other urban centres suggest, the main towns were bursting with fruit and vegetables. The Arcadian experiment was paying off – at least for those with access to land. Property owners either worked their own land or hired gardeners to do it; the poor, if they worked hard, could in theory rent enough land to keep themselves in food until they had amassed sufficient savings to purchase their own plot. But for many, the reality was different.

Chinese gardens

With the discovery of gold in the 1860s, Chinese men began to arrive in New Zealand. Some came from the Australian goldfields, and before long others were coming direct from China. Some struck gold, but many realised the opportunity that lay with a captive audience who needed to be fed. Market gardening quickly became synonymous with the Chinese community in New Zealand – a story that has been well told.

With Chinese gardens came Chinese vegetables, Chinese traditions of vegetable production, and a somewhat uncomfortable symbiotic relationship with the European population. For the European market,

Chinese gardeners grew and sold familiar vegetables such as potatoes, corn, cabbage, peas, turnips, gooseberries and strawberries, as Boileau has recorded; one Dunedin article from 1878 lists rhubarb, lettuces, celery and cauliflowers in the garden of Leong Foy of Caversham.[98] They also grew vegetables for their own consumption, such as Chinese radish, Chinese parsley, garlic, celery and Chinese cabbage, which were suited to the cool temperatures of New Zealand.[99] This array of vegetables ultimately found its way into the horticultural palette of Pākehā society.

It was generally understood that Chinese gardeners relied on manure – both animal and human – to maintain soil fertility. Lily Lee and Ruth Lam suggest it was highly unlikely that Chinese market gardeners used 'night soil' (i.e. the contents of a bucket toilet) in commercial gardening in New Zealand, but write that it was 'not unheard of in' Chinese home gardens.[100] This point has been recently emphasised by James Beattie and Joanna Boileau.[101] A first-hand account exists from Round Hill in Southland, where a significant Chinese township developed in the 1880s in an area that was mined by Chinese through to 1907. Market gardeners there were observed using urine collected from the local school in direct application on leafy green crops. Whether night soil was used as well is not known.[102]

Chinese garden fertilisation methods were a point of tension for European settlers, who were already struggling with effective disposal of human waste in the growing cities. Letters to the editor on the topic appeared from the 1870s. 'Watchman' wrote to Dunedin's *Evening Standard* in 1878, observing that the city council were 'at their wits' end to tell what to do with the City sewage'. Speaking as someone with 'knowledge of the way the Chinese manure their market gardens, and the way disease is carried from house to house', Watchman sincerely hoped the council would not let Chinese gardeners make use of the waste.[103] Some went so far as to link Chinese gardening methods with typhoid, due to what one writer described as their 'filthy mode of conducting their agricultural operations'.[104] Perhaps, however, as an Ashburton correspondent noted in 1910, this prejudice existed 'just because they are foreigners in competition with producers of our own race'.[105]

Fertilisation with human waste presented a cultural challenge for Māori too, who would not use excrement of any kind in their gardens.

Boileau has posited this as a reason for the decline of Māori market gardening: as their expanded production exhausted soils, the traditional method of burning new garden sites became untenable.[106]

Newspapers from the period mention a surprising number of thefts from Chinese gardens by Europeans.[107] This practice continued well into the 1930s, causing one magistrate to comment in 1935, 'Some people think it is not theft but a joke to steal from Chinese gardens; that view has to be corrected.'[108] Some Chinese gardeners were also victims of violent attacks, possibly as a result of a more general anti-Chinese campaign towards the end of the century.[109] It is difficult to draw any conclusions from these episodes: in the minds of some, perhaps, Chinese were not viewed as the legitimate owners of their gardens – at least, not in the same way that a Pākehā gardener owned his or her patch. Certainly, a view prevailed among many Pākehā market gardeners that their Chinese rivals were driving them out of business.

The destruction of Māori gardens

The story of gardening was completely different for Māori. War over land sales broke out in the Bay of Islands in 1845 between Ngāpuhi and government troops, the first of a series of conflicts in the North Island now referred to as the New Zealand Wars. The conflicts were hugely disruptive and had an enduring impact on Māori life and gardening. In the 1850s, in an attempt to halt land sales, the Māori King movement Te Kīngitanga began. As the movement gained traction, Māori became less inclined to sell land, and tensions between Europeans and Māori escalated.[110] In addition, as historian Harry Evison has pointed out, European opinions of 'natives' – indigenous peoples within the lands of the British Empire – turned sharply after the so-called 'Indian Mutiny' of 1857. From this point, 'all agreed that "natives" must be firmly governed, and (if possible) civilised'.[111] The prevailing view in New Zealand was that Māori communal ownership of land was the main impediment to their becoming 'civilised', and a process of privatising Māori land into separate allotments commenced.[112]

With this convergence of ideas at play, the 1860s–1880s marked the destruction of Māori gardening activities in many areas. Just 50 years earlier Māori had formed the majority of the population. Now they were in the minority. They had lost access to mahinga kai and food-growing areas, both through land deals that ultimately ignored provisions for protecting important sites, and as a direct result of the Wars. From 1863, for example, the many communal gardens dotted along the banks of the Waikato River disappeared. Yet some Māori – like Hare Puke (Ngāti Wairere, born in 1889), about whom we shall soon hear more – were still taught the traditional horticultural methods and grew up gardening at home with their whānau.[113]

Many descendants of Turi – mentioned in Chapter One for establishing Taranaki's first kūmara gardens hundreds of years earlier – found themselves landless as a result of phoney land deals made with settlers. Te Matoe, also called Te Karere Omahuru, petitioned the government for redress throughout the 1850s and into the 1870s, without success. His six-year-old son, Ngātau Omahuru, was kidnapped in 1868 by colonial forces and ultimately adopted by Premier William Fox and his wife Sarah. Many of Te Matoe's family were members of the Parihaka community.[114] Their gardening story is also explored further in this book.

Just as they would later from Chinese gardens, Europeans were known to help themselves to – or steal – produce from Māori gardens early in the century, and in some places the practice continued. In 1864 the resident magistrate in Lyttelton, Dr William Donald, wrote to the superintendent requesting that signs be put up at the entrance to the kāinga at Rāpaki 'cautioning trespassers', because Europeans were constantly plundering the garden.[115]

As more and more Māori were driven off the land through legislative, military or other means, access to their abandoned gardens proved convenient for Europeans moving through the countryside. In his recollections of travelling through the North Island in the 1860s and 70s, Frederick Spencer described the convenience of gathering food in places from which the inhabitants had been displaced: 'In those days, if passing through what had been or was a Māori village, there was generally plenty of fruit in the season.'[116] On one occasion he was told of 'a lot of apples going to waste at an old Maori Village ... sure enough, there were hundred weights all ripe, and pigs enjoying themselves'. In 1875 he recorded finding bunches of grapes hanging from a vine in an abandoned village. Spencer's father, a missionary, recalled laden peach trees in Māori villages.[117] Spencer was witnessing the aftermath of the destruction of Māori communities, a direct result of colonisation.

A 2009 report from the Waitangi Tribunal detailing the campaign in the Urewera area in the late 1860s describes one example of the destruction of Māori gardens by government forces. The government had confiscated Tūhoe lands in 1866, including three 'large kumara and potato gardens'. In the northern part of Tūhoe territory, where these gardens were sited, could be found plantations of kūmara and harakeke described as 'crucial to the overall integrity of the Tuhoe economy'.[118] Dispossessed of their lands, the Tūhoe people retreated into the interior where they established large potato gardens to ensure their survival in this time of war. In an attempt to wipe out Tūhoe's economic base, the colonial forces set out to destroy these gardens. Colonel Whitmore, who by 1869 was essentially commander-in-chief of the army, led seven distinct campaigns as part of the New Zealand Wars. Perhaps the most challenging of these was the campaign in the Urewera. Whitmore reported in 1869 that 'the main object' of this expedition 'was the destruction of the food supply in the mountains':

> Every potato garden was securely fenced, and the removal of a
> small portion of the enclosure ensured the co-operation of the
> troops of pigs running in the forest … the whole of the ground was
> turned over, and all the potatoes they did not devour were spoiled
> by the heavy frosts, which occurred every night.[119]

Those Urewera hapū who surrendered, such as Ngāti Whare, were shifted to reservation land on the coast in an attempt to get them 'out of their mountain fastnesses', as Defence Minister Donald McLean put it in 1870.[120] They were given seed potatoes, but those planted in the new reserve failed that year due to the dry season and a lack of adequate tools. The experiment resulted in considerable misery.[121]

Worse than the destruction of those gardens in 1869, however, were the government depredations in 1870 at Lake Waikaremoana. Here government forces found an enormous quantity of potatoes growing, described as 'something marvellous', an 'almost fabulous amount of potatoes', enough to 'keep 1000 men for fifteen months'.[122] During June 1870 the soldiers destroyed around 100 cultivated areas: 'we … estimate the quantity already taken, eaten or destroyed at (including a large quantity of seed potatoes) certainly not less than 200 tons'.[123] The Waitangi Tribunal has found this destruction 'served no useful strategic purpose'.[124] In 1871 as the campaign in this area continued, the government forces relied on Māori cultivations, which included 'newly planted potatoes, peaches and apples'.[125]

By the time European colonists began arriving in Aotearoa, Māori were growing many food crops that were familiar to the new settlers. These food supplies became an integral part of the economic exchange network that developed between Māori and the colonists, and made the process of European settlement more manageable than it would otherwise have been. One could say that, to some degree, Māori food gardens actually enabled colonisation.

Perhaps Europeans were unaware of the negative impact their activities had on the indigenous peoples who had been in Aotearoa for hundreds of years. Settler families established themselves on their newly purchased land and planted their own food crops, after which they no longer relied on Māori gardeners for staple foods. Through land sales and military intervention, Māori were increasingly excluded from traditional food-gathering and growing areas. The result was significant suffering and hardship among many Māori communities.

CHAPTER 3

The City Beautiful and Garden City movements: 1900–40

Few colonial gardeners had the time and resources to beautify their property with flowers and shrubs, so focused were they initially on food production. Some were inspired to beautify, of course, and many brought with them a well-developed love of gardening. Trader Thomas McDonnell established an ornamental garden in the Hokianga during the 1830s and 40s which he stocked with rarities from India, China and everywhere else, sourced directly from the Royal Botanic Gardens in Kew.[1] Sarah Mathew nurtured flowering bulbs in her Auckland garden in 1840, and in 1842 oversaw the establishment of a garden with lawns, flowerbeds, 'a belt of shrubbery partly native trees', and planted with oaks, chestnuts, walnuts and 'vines and fig trees' imported from Sydney.[2] Alice Porter of Auckland wrote to her brother in Nelson in 1843 about laburnums and her search for a cutting of a Damask rose, which she prized 'very much'.[3]

But beautification was not the primary emphasis of most European gardeners in New Zealand. Mary Hobhouse, wife of Bishop Edmund Hobhouse, complained from Nelson in 1860 that her garden consisted of

> a bit of rough lawn adorned by one solitary round bed ... and
> divided from the kitchen garden ... by a hedge of roses and common
> red geranium. The remainder is a square of open border filled with
> a miscellany of flowers and shrubs of which it is hopeless to make
> anything ornamental ...[4]

In 1861 she wrote that a flower garden was 'a thing absolutely unknown here'.[5] She decided to content herself with a few common flowers, and concentrated her efforts 'on the *tidiness* of the garden a point sorely neglected generally'. Her husband's cousin had 'made his little garden a perfect model', but he, she felt, was an exception.[6]

Horticultural societies around the country promoted flowers at their shows with limited success. In 1873 the Auckland Horticultural Society was 'reluctantly compelled to admit that a considerable amount of apathy exists amongst a large section of the community with regard to Horticultural pursuits'.[7]

At the dawn of the twentieth century, however, interest in gardening for aesthetic reasons rather than food production finally began to take hold. These are the years we must look to for the beginning of 'City Beautiful' and 'Garden City' thinking in Aotearoa. The City Beautiful movement was primarily an American move to reform degraded urban environments by introducing grand scenes and more (and bigger) green spaces into them. Washington DC stands out as an early (1902) exemplar of this reform approach, which swept through US cities and beyond; the plan for the Australian city of Canberra was also heavily influenced by City Beautiful thinking.

From food to decorative features

In New Zealand, fledgling attempts to improve the visual appearance of urban spaces began with the formation of the Dunedin Amenities Society in 1887, followed by the Christchurch Beautifying Association in 1897. Thereafter other groups – among them the Timaru Beautifying Association (1889) and the Wadestown Cottage Garden and Beautifying Society (1917) – sprang up around New Zealand.[8] In Christchurch, where the horticultural society had barely staggered through the 1890s, from 1903 – with the help of the rose and chrysanthemum societies and their focus on beautification – it was up and running as the Canterbury United Horticultural Society.[9] Within a year its membership climbed to 160, and by 1907 it stood at 230.[10]

The idea of urban renewal through beautification was pertinent to many in New Zealand towns by the turn of the century. Crowded infill housing in some working-class areas, industrial pollution and issues with the disposal of sewage and other waste meant some urban areas were little better than slums.[11] Chinese gardeners' use of human waste in liquid manures was an easy target under the circumstances, and a bubonic plague scare in 1900 elicited comparisons between Christchurch and Samuel Pepys' seventeenth-century London.[12] From the 1870s at least, commentators had touted the benefits of trees and other plants in ameliorating unhealthy urban spaces.[13] Artist Alfred Sharpe argued for the improvement of public gardens in Auckland in 1880, in part for health reasons.[14] Efforts by town planners to address urban health problems by incorporating ideas of town planning in their designs began to have an appreciable impact from the late nineteenth century, and was further developed by new 'garden city thinking' in the early twentieth century. New suburban developments in South Dunedin reflected this, as did the Clifton 'garden suburb' development in one of Christchurch's hill areas from 1902.[15] Decorative garden-making was one means of transforming these spaces, and the idea caught on.

David Hill has claimed that Lower Hutt was the first 'garden city' on the basis of its relatively large (quarter-acre) private sections.[16] The Taranaki area, regarded by Māori as 'the garden of the country' (as noted by Wakefield in the 1830s), was referred to as 'the garden of New Zealand' around the 1890s.[17] In 1900 the *Press* described 'garden-loving Christchurch', and Christchurch appears to have adopted the title of 'Garden City' by 1906, even if it did not catch on for a decade and a half.[18] In 1924 the mayor of Christchurch claimed the town was 'generally acknowledged' as the 'Garden City of the Dominion'.[19] Councillor Andrews, chair of the Reserves Committee, praised the beauty of the city's parks and 'paid tribute to the manner in which the residents looked after their gardens, and thus assisted to make Christchurch so beautiful'.[20]

An example of one early gardener's effort to beautify their frontage is a Merivale cottage garden on what was Boundary Road in the 1890s. Lilies lined the shell or gravel path to a verandah over which climbed a creeper, while young cabbage trees were planted in the coarse lawn. In Sydenham, one Johnson Street garden had shell paths and box hedges.[21] Front gardens could look elegant.

Newspaper gardening writers began to devote an increasing amount of space to decorative planting. In 1893 the *Star* began a column that had a definite bias towards ornamental gardening: on 20 May 1893, for

An elegant front garden on Boundary Road, Christchurch, 1890s.
Photo courtesy of Pip Middleton

example, the kitchen garden section simply mentioned the need for trimming hedges, forcing sea kale and protecting rhubarb. In contrast, there was much to be attended to in the flower garden: shrubs should be pruned, borders tidied, leaves gathered to decompose behind the shrubbery, and hyacinths, crocuses, snowdrops and perhaps narcissus and anemones could now be planted. Peonies were discussed alongside the importance of deep trenching in preparation for rose planting. Greenhouse operations were equally detailed: chrysanthemums needed attention; primulas would benefit from a mix of leaf mould, sand and old cow manure; Christmas roses should be potted; and clumps of *Schizostylis coccinea* and *Lycoris radiata* could be 'lifted from the open border and transferred to the stage in the greenhouse'.[22] Roses and chrysanthemums began to figure prominently: the *Press* dedicated an entire column to a report on the 1895 chrysanthemum show.[23] Clearly, flower gardening was becoming popular.

The Garden City concept

In 1898 Ebenezer Howard published his text on town planning in Britain, *Tomorrow: A peaceful path to real reform* (reissued in 1902 as *Garden Cities of Tomorrow*). In it Howard set out the main tenets of his concept for

Flower Garden, Auckland *by Robert Walrond (c. 1913).*
Museum of New Zealand Te Papa Tongarewa, A.018186

the 'Garden City', which would be designed to capture the best aspects of urban and rural living with none of the associated problems. 'Human society and the beauty of nature are meant to be enjoyed together,' he stated:

> As man and woman by their varied gifts and faculties supplement each other, so should town and country … Town and country *must be married*, and out of this joyous union will spring a new hope, a new life, a new civilization.[24]

Howard's ideal Garden City was laid out in a concentric pattern of approximately 6000 acres (2500ha) with a circular 1000-acre urban area in the centre. In the middle of this was a park of 5.5 acres (2.2ha) 'laid out as a beautiful and well-watered garden' and surrounded by public buildings. These, in turn, were surrounded by private dwellings, with the industrial areas in the outer ring. Ideally the town would cater for around 30,000 people, with another 2000 on the surrounding agricultural estate.[25] A Garden City was the ideal size for walkability and commercial viability, and would ensure an answer to 'the problem of how to restore the people to the land … the very embodiment of Divine love for man'.[26] Bourneville in Birmingham, England, a model village enshrining some of these values, had been established between 1893 and 1900, and in 1904 Howard's Garden City Association built Letchworth Garden City in Hertfordshire, England.

The *Press* reported in 1900:

> One of the newest associations in London appears under the name of 'The Garden City Association'. Its object is to induce manufacturers, and employers of labour on any large scale, to remove from cities, and set up their works in country districts now nearly deserted – thus reviv-ifying the sparsely populated agricultural areas and lessening the 'drift towards towns' ... Mr Ebenezer Howard's really remarkable book ... carried the idea into detail with such success that the Association now formed resolves to work exactly on the lines which he suggests ...[27]

According to Howard, better planning would place the needs and well-being of people, particularly those of the working class, at the forefront of development. His contribution to the town-planning movement in New Zealand has been widely acknowledged, and the Housing and Town Planning Act 1909 can largely be attributed to his efforts. His utopian Garden City concept – and the notion of town planning – was a radical departure from the way towns and cities had developed historically in Britain, and an example of the periphery affecting the metropolis.

The potential of planned towns had already been demonstrated in the colonies. Planned towns were built in the American colonies during the seventeenth century, and the Australian towns of Adelaide (1836) and Melbourne (1837) demonstrated the efficiency of the grid layout as an organising principle. Canberra is an excellent example of a planned town and was under development from 1908. As these cities were to be a model of how Britain could be reinvented, it is no wonder that colonials took an interest.

Howard's ideas were warmly received in Christchurch, in particular by the town beautifiers. In 1900 the *Press* reported on Howard's progress:

> A plan for 'Garden City' is ... already drawn, and we hear that 'ideal-ists are enchanted by the bright vista of boulevards, gardens, model buildings, chaste architecture, and other pleasing characteristics', while the social reformer hails the scheme as a possible relief from the growing and terrible evils produced by overcrowding.[28]

From 1891 to 1894, British lawyer and politician Sir John Eldon Gorst was part of a Royal Commission on labour and involved in an inquiry into the Poor Law schools, or workhouses, of London.[29] Malnutrition was rife in these institutions, and his concern with 'the physical degeneracy amongst British school children resulting from wrong-feeding' was reported in the *Press* in 1904.[30] In 1906 Gorst published *The Children of the Nation: How their health and vigour should be promoted by the state*, a publication that is thought to have foreshadowed the principles of health reformer and founder of the Plunket Society, Truby King.[31] In it Gorst examined the advantages of Letchworth:

> The Garden City at Letchworth, in Hertfordshire, holds out advantages to both employers and employed … Every cottage will have a garden or an allotment within easy reach … The streets of the town will be broad avenues planted with trees, letting light and air into the heart of the city, and there will be parks, playgrounds, and open spaces, so as to make the place beautiful as well as healthy. In this city the worker will have a healthy home, and his wife and children will live in conditions nearly approaching those of country life.[32]

Gorst was also enthusiastic about Bourneville: 'Every house has its garden, by no means restricted to the growth of saleable produce. There are luxuriant flowers in front of each dwelling, as well as useful fruits and vegetables behind …'[33]

Howard's Garden City Association staged a display at the New Zealand International Exhibition, which opened in Christchurch at Hagley Park on 1 November 1906. The association's message was 'to promote the relief of overcrowded areas, and to secure a wider distribution of population over the land'. Its Garden Cities were 'designed to secure healthful and adequate housing, in which the inhabitants shall become in a collective capacity the owners of the sites'. The association advocated for 'the removal of manufactures from congested centres to the country' and to improve 'the conditions of existing towns'.[34] The display included books on the subject, a plan of Letchworth and images of experimental townships such as Port Sunlight, a model village in Cheshire.

Gorst was also present at the exhibition as a special envoy of the British government. He was so impressed with what he saw in Christchurch – the free access of its inhabitants to parks and gardens, green spaces and fresh air – that he was moved to compare the city to the Letchworth experiment. Sir John Hall, who in the 1850s had championed the beautification of the young Canterbury settlement, was wheeled out as honorary mayor for the duration of the exhibition, and was no doubt proud to see his city showcased in this way.

The Garden City concept proved a great enabling myth for the beautifiers and found purchase in Aotearoa, in particular in Christchurch. There is a subtle but important distinction to be drawn, however, between the view of Christchurch as a 'pretty' place, and Christchurch as a wholesome environment for people of all walks of life to live in. The latter interpretation is likely what Gorst would have meant in his oft-quoted remarks.

Although the beautification agenda was in large part aimed at the labouring classes, it was policed by the middle classes, as we shall see.

Cottage garden competitions

The 1906–07 New Zealand International Exhibition ran for five months, during which time the Canterbury United Horticultural Society held four major flower shows. The *Star* garden columnist advised of these as early as July 1906, and suggested that gardeners should lose no time 'in making preparations if they wish to win prizes'.[35] According to James Cowan, author of *The Official Record of the New Zealand International Exhibition of Arts and Industries*, the shows were 'the most beautiful and comprehensive yet organized in New Zealand'.[36] The dahlia display alone, with 265 entries, required 304 metres of tables and an additional 213 metres of floor space. New Zealand native plants also featured.[37] Christchurch, according to Cowan, was 'a city of flowers' – exactly the perception the beautifiers wished for.[38]

Horticultural societies all over the country were beginning to hold competitions directed at the labouring classes – the 'cottagers' – in an attempt to encourage the owners of smaller properties to improve their gardens. The earliest was possibly a cottage garden competition held in Wellington in 1842.[39] Dunedin Horticultural Society held a cottage garden competition in 1898 that was not well supported; it fared a little better in 1899, however, with eight entries.[40] The report in the *Otago Witness* clarifies the term 'cottage' garden: it was 'to include fruit, flowers, and vegetables, and in area was not to exceed ¼ acre [0.1ha] (house included)'. One outstanding Dunedin entry was a plot of just one-eighth of an acre. The garden was commended for 'its clean asphalt walks, a perfect little lawn with trees loaded with fruit, a good variety of flowers (including a magnolia in bloom in a pot), and vegetables good in variety and quality – everything was neat and well kept ... It could hardly be realised that such a calm beauty spot could be found right in the heart of the city.'[41]

Bruce Horticultural Society in Otago held a cottage garden competition in 1902; in 1904 it held separate cottage garden and flower garden competitions, suggesting that the cottage garden was perhaps more utilitarian.[42] Kaitangata, also in Otago, held a cottage garden competition in conjunction with a horticultural show in 1909; the latter awarded prizes for vegetables only.[43]

By 1900, to enable the gardener to tend to all of this activity, a growing array of gardening tools was available in New Zealand. Dutch hoes, Warren hoes and ordinary hoes, hedge shears, border shears, pruning shears, budding knives, vine pruners, vine scissors, edging knives, syringes and 'ladies' garden sets' could all be purchased from John

Parkhurst, Woodbury, in the 1890s: the residence of Horace Edgar Musgrave. Note the rough scythed lawn.

Photo courtesy of David Musgrave

Duthie & Co in Wellington.[44] The specific uses of these tools says much about the horticultural knowledge of the period.

Around this time lawns began to attract attention. In 1861 settler Mary Hobhouse wrote that a cow was just as effective as men with scythes for keeping a lawn down; by 1900 gardeners could choose a lawnmower from a range of brands including Champion, Excelsior, New Easy and Anglo-Paris.[45]

The Homestead Beautiful

The *Journal of Agriculture* in 1913 promoted the idea of 'the Homestead Beautiful', which encouraged rural folk to develop elegant garden designs that incorporated a graceful driveway, large orchards, vegetable gardens and lawns with shrubs and specimen trees, for 'a clean, wholesome and attractive' appearance.[46]

The diary and photographs of William Turton, owner of Waihi Bush farm near Geraldine, demonstrate how new approaches to landscaping and gardening could play out in a rural context. Theirs was very much a 'Homestead Beautiful'. A 1926 photograph of the grounds shows that Turton and his wife Marion were influenced by *japonisme*, the study of Japanese art.[47] The shingle drive was kept in immaculate condition and elegant cabbage trees framed various views. The house was offset by neat gardens of shrubs and flowers, and draped in japonica and wisteria vines so vigorous they had needed a serious pruning in March 1922.[48] The vegetable gardens required considerable upkeep that year: flower borders needed 'doing up' in January, the orchard required 'spraying mixture' in July, and manure was applied to the gardens in August. But the feature that received the most attention by far was the expansive lawn.[49] In 1904 the farmhands and gardeners cut, cleared and rolled the grass frequently; they were still working on it in 1922, when farmhands and Turton himself all took part in cutting and raking.[50] '[S]pent a lot of time over the lawn mower,' Turton lamented in 1923.[51] It was expensive and time-consuming work: one farmhand spent five consecutive days mowing solidly in December 1924.[52]

The rise of the ornamental garden

In the early 1900s orchards and vegetable gardens still figured in advertisements, but increasingly, lawns, shrubs and flowers received special mention. A Remuera 'gentleman's residence' had '[b]eautiful grounds ...

Marion Turton in her garden at Waihi Bush, 1926.
Photo courtesy of David Musgrave

Marion (left) and William Turton, Waihi Bush, 1916.
Photo courtesy of David Musgrave

Marion and William Turton in their beloved car at Waihi Bush, 1920s.

Photo courtesy of David Musgrave

Flower Garden, Auckland *by Robert Walrond (1915).*

Museum of New Zealand Te Papa Tongarewa, A.018195

shrubs, flower-beds, drives, tennis court', and an Epsom bungalow of 1919 had 'flower and vegetable garden, lawn'.[53] One gardening column-ist claimed that a 'well-made and properly kept' lawn was 'one of the most important features in connection with any garden, and especially that surrounding the residence'.[54] According to the *New Zealand Herald*, 'nothing … adds more to the beauty of the homestead than well-kept lawns and grass plots'.[55]

A 3–4-acre (1.2–1.6ha) property at Lake Takapuna in Auckland, advertised in 1909, claimed to be 'beautifully laid out in croquet lawn (with tea pavilion), flower garden, kitchen garden, and paddock. Paths asphalted and surface tile drained. The garden is all thoroughly spade trenched, and planted with Choice Varieties of Bulbs, Flowering Plants and Shrubs.'[56] In Lower Hutt in 1910 one home was described as hav-ing 'well-grown shelter trees, ornamental, orange, and lemon trees … [and] a small glass-house'; another had 'all classes of trees and flower-ing shrubs; strawberries and other small fruits'.[57] In Auckland in 1911, an Epsom house was offered for sale with 'gardens, lawns, orchard … magnificent trees'; another had a 'beautifully laid out lawn, fruit trees, vegetable garden, greenhouse'; and a cottage garden in Ellerslie boasted 'rich volcanic soil: fruit trees, vines, lawn, and vegetable garden'.[58] Yet another advertisement described a garden with shrubs, flower gardens, large lawns, palms and even rare and native plants.[59] Some of these gar-dens may have been laid out and maintained by landscape gardeners who benefited from the rise of the middle classes – as was the case for landscape gardener Alfred Buxton in Christchurch.[60]

Examples such as these abound, and many garden historians have considered them indicative of fashions of the time. There are impor-tant differences between fashion and common practice, however. The diary of Christchurch man Harry Wills provides a fabulous window into the garden of an ordinary householder of the late 1920s and early 1930s. In some ways his garden work was similar to that of William Trotter in the 1850s (mentioned in Chapter Two): with the produce harvested from his own back yard and contributions from friends, Wills was largely self-sufficient in fruit and vegetables. Like Trotter, he poured a lot of energy into food production. Typical diary entries include: 'planted 2 rows carrots 1 Parsnips, 4 Onions, 3 Pease some lettuce & radish & 36 potatoes'; 'hoed the garden … planted 16 goose-

The Ross family in their garden walk, Christchurch, 1914.
Photo courtesy of David Musgrave

berry cuttings'; 'Got 4 Doz. Tomato plants ... & some Scarlet runners'; 'picked raspberries & peas'.[61] Unike Trotter, however, Wills expended a lot of energy on maintaining a lawn and establishing a flower garden: 'did some digging & cut the lawn' is an oft-repeated phrase.[62]

Audrey Potter described a similar garden that her parents established in 1929. Situated in Riccarton, Christchurch, it was about half an acre (0.2ha) in size and combined the functional with the ornamental. 'A good part of it was put down in orchard': four plum trees, a Lord Wolseley apple, a Granny Smith, a Cox's Orange, a Sturmer, a Delicious – 'which was nearly always a red one in those days' – and 'a huge, very round, very dark red, almost black apple', perhaps Black Prince.[63] There were also two walnuts, two pears, a nectarine and a white-fleshed peach. A prolific vegetable garden produced potatoes, peas, carrots, parsnips, cabbages, silverbeet, beans and pumpkins and sweetcorn.[64]

The ornamental plantings included a rose bed, lilac, a rhododendron, a flowering cherry, geraniums and hydrangeas, forsythia and a Judas tree. Grass in the orchard was kept in check with a scythe, while the lawn proper was cut by the children using one of two push mowers: 'One was easier to push than the other. That was the one I usually got,' Audrey recalled. With no man in the house, her mother maintained the garden with assistance from Audrey and her siblings and their grandfather, who lived nearby. According to Audrey, 'the vegetable garden … was really more important than the flower garden'.[65]

In comparing this typical garden with a more formal garden of the same period, we can see the same ideals at play, albeit on a larger and more intensive scale. June Stewart remembered her parents' prize-winning Papanui garden of the 1920s: 'Mostly it was large areas of lawn, and there was a big orchard down the back that was covered in daffodils in the spring.'[66] It was something of a 'wild garden', with bamboos, mature trees and shrubs, and had large asparagus beds and 'an enormous vegetable garden out the back'. The orchard contained apples (Gravensteins, a Black Prince and Sturmers), pears (Winter Cole and Winter Nelis), quinces, and a row of peaches that ripened at different times. There were red and white currants, a large walnut tree, and cherries in 'a big cherry house' that also accommodated raspberries.[67]

Home garden competitions: 1917–1930s

While we might assume that attention to flower gardens would have dropped off during World War One (1914–18), interest in garden societies actually seems to have picked up at that time. In a sign of the growing popularity of one flower, the first specialist rose business in New Zealand was established in Feilding in 1912.[68] The Otago Women's Club Gardening Circle held regular flower shows in 1915, and its 1916 show featured chrysanthemums, hydrangeas, carnations, roses and heliotropes.[69] Greymouth Horticultural Society held a Sweet Pea and Rose Show for the first time in 1916, and the *Grey River Argus* subsequently claimed that a 'number of well-kept gardens' had 'sprung into being as a result of the interest the Society's show has created'.[70]

The Otago Women's Club Gardening Circle, which had hitherto been strictly focused on flowers, in mid-1917 began to include fruit and vegetables, samples of which were distributed at one meeting.[71] Wellington's Wadestown Cottage Garden and Beautifying Society, established during World War One, showed an initial preference for vegetable and fruit

View of Clevedon Road, Ōtāhuhu, 1910. Note the extensive vegetable garden, orchard and chicken house.

Auckland Libraries Heritage Collections, Footprints 04297

gardening. At its first annual meeting in May 1918 the group agreed 'to help any member, who may be called up for Service, with regards to keeping his family in vegetables while away'.[72] A later meeting was postponed to allow for a public demonstration of fruit tree pruning and a talk about vegetable growing, and the Department of Agriculture promised to give members fruit trees, providing they looked after them.[73] In 1919 the Wadestown society was focused on roses, chrysanthemums and rockeries, and its horticultural shows included classes for vegetables, flowers and, in 1921, fruit as well.[74]

From 1917, apart from a hiatus in 1922–24, regular home garden competitions were held in Christchurch.[75] As well as small local events, at least three organisations – the Sweet Pea and Carnation Society, the Canterbury Horticultural Society and the Christchurch Beautifying Association – held citywide contests. The focus of these was still fixed on 'cottagers': the 1918 Canterbury Horticultural Society competition was a 'cottage garden' competition, although the prize schedule shows that the expected layout of the gardens was not the intermingled vegetables, herbs and flowers that Christine Dann has spoken of in her book on cottage gardening, but instead were likely to be typical workers' gardens – small cottages on modest sections, with ornamentals out the front and vegetables out the back.[76] Although initially minimal, the growing number of entries in the various competitions indicates that public interest increased from the late 1920s.

Wadestown Cottage Garden and Beautifying Society also held home garden competitions. This organisation placed equal value on flowers and vegetables, and when it came time to judge the best-kept gardens in 1919, a separate judge was appointed for each section.[77] The Runanga State Collieries Horticultural Society near Greymouth held a similar event in 1920 and offered prizes for a 'Collection of vegetables and rotation of crops: collection of flowering plants: Perenails [sic], Herbacious, and Annuals: Collection of Fruit trees, neatness and general layout of the whole garden'.[78]

Beautifying societies also put energy into making public places attractive. In 1918 the *Press* reported on a visit by members of the Christchurch Beautifying Association (most of whose efforts reflected an interest in beautifying public sites, including railway stations and school gardens) to the Woolston Tanneries 'to inspect the gardens laid out on part of the factory site'. Club members were impressed and remarked that they had 'no idea, no conception, that the somewhat despised borough of Woolston – despised as regards beautification – possessed in its centre such a beautiful spot'.[79] The Timaru Beautifying Association concentrated its early efforts on planting trees and establishing flowerbeds in public spaces to 'remove an eyesore of the public'.[80] In Dunedin in the mid-1920s the gardening circle supplied New Zealand Railways with wallflowers, roses, lobelia, asters, stocks, snapdragons and other plants for its various stations.[81]

A 'source of spiritual uplift'

Membership of the Canterbury Horticultural Society climbed steeply in the 1920s and 30s. The society had 537 members in 1927, 773 in 1928 and welcomed its 1000th member in 1929.[82] It is striking that this spike in membership occurred during the Depression. In a letter to her publisher in 1932 poet Ursula Bethell wrote, 'Many confess that they are surprised to find themselves *happier* since the Slump ... A rose-nurseryman whom I visited the other day ... said he hadn't done at all so badly this year. People stopping at home more, & working in their gardens ...'[83] Beautiful gardens, the *Press* suggested, could be places of respite, a 'source of spiritual uplift'.[84]

These years also coincided with a rise in entries for horticultural shows: 830 for the society's 1934 gladiolus and dahlia show for example, compared with 595 the previous year.[85] The Canterbury Commercial Travellers and Warehousemen's Association also held home garden competitions at this time, and entries in 1934 were the highest since competitions began in 1930. The prize list included awards for flowers, a vegetable garden, a fernery and lawns.[86] The bubble was short-lived. From 1935, interest in Christchurch garden competitions dropped to fewer than 100 entries; from 1944, even the largest competition had fewer than 50 entries.

Given this brief burst of popularity, one might think that garden competitions had a limited impact in the city, but this is not the case. Like garden-produce exhibitions, home garden competitions modelled a certain kind of righteousness. They established a criteria for what was considered 'worthy', and publicly rewarded those who followed the prescription. There was a certain amount of pressure on others to follow suit. The *Lyttelton Times* outlined the effect of the competitions in 1925:

> While the average amateur gardener is convinced he can never do
> as well, he is generally impelled to do a little better than he has
> been doing ... Even to read about those paragon plots is enough to
> impress the generality of mortals with a sense of guilt or shortcom-
> ing, as though perfection in domestic horticulture were one of the
> essential virtues.[87]

Care for gardens, according to the *Times*, implied 'the love of home and a strong sense of the beautiful, sentiments that lie at the root of real civic greatness'. It more than justified the continued existence of the Horticultural Society.

Jack Humm echoed these sentiments in 1930 in his report on gardens in the suburbs of Sumner and Redcliffs: 'These prize gardens set a standard in horticulture, and show what can be accomplished in garden making. Competition also urges those interested to greater effort, and encourages the residents to take greater pride in their home grounds.'[88]

Otahuna, the residence of Sir Heaton Rhodes, president of the Canterbury Horticultural Society from 1898 until his death in 1956, was an example of the 'prize gardens' to which Humm referred.[89] Rhodes was hard-working and generous, and his garden demonstrated the potential for magnificence in planned cultivation. Otahuna was, and still is, famous for its stunning daffodil displays, which Rhodes shared with the people of Christchurch. His love of growing bulbs was apparently sparked in Yokohama, Japan, where it seems he purchased some lily bulbs in 1891. According to Geoffrey Rice, by the end of the 1930s Rhodes and his daffodils were 'firmly fixed in the public mind as a key icon of Christchurch's self-image as "the Garden City"'.[90]

While competitions and properties like Otahuna established or reinforced a kind of moral code about gardening, they did not necessarily influence the kinds of gardens people made. Ordinary gardens were still food gardens, but increasingly they were enhanced with flowers, shrubs and well-trimmed lawns. William and Minnie Basset's garden in North Parade, Christchurch, exemplifies this. The Bassets settled there on arrival from Ireland prior to World War One. At the back of the house they established a garden divided by brick and cinder paths and containing every imaginable vegetable, as well as raspberries, strawberries, Jerusalem artichokes, plums, apples and peaches. The sides of the property were shielded by laurel and macrocarpa hedges, and the front garden consisted of a lawn with shrubs along the low front fence and a formal flowerbed in front of the living room window.[91]

Adam and Mary Fraser's garden in Ayr Street, Mosgiel, west of Dunedin, also demonstrates this combination. In the 1940s the 'huge' front garden, Mary's 'pride and joy', was a blaze of colour. The lawn was 'always immaculate' and, according to her granddaughter, too good to play on. Adam's garden, around the back, was full of vegetables and fruit, with raspberries on hooped frames, cape gooseberries and a large stand of blackcurrants. He grew all the usual vegetables, and plaited his onions tidily before hanging them in the shed.[92]

The Masterton garden of Bert and Annie Wallis is another example of this style of garden. Bert and Annie arrived in New Zealand from the United Kingdom in the 1920s as young adults with a passion for bearded and species irises, of which they grew many. They had two rock gardens – a sunny one with dwarf tulips and a shady one with cyclamens and trout lilies – and a pond surrounded by Japanese irises, peonies, roses and shrubs.[93] The garden included 'a large *Viburnum rhytidophyllum*

The pond in Bert and Annie Wallis's garden, Masterton, 1960s.
Photo courtesy of Gareth Winter

… a Rhododendron Grande, a double clematis, a wisteria-encrusted aviary', a glasshouse full of orchids, and a shade house for frost-tender crested irises and ferns. Bert also maintained a large vegetable garden and kept a small orchard along the side of the house. Bert and Annie were founders of the New Zealand Iris Society and 'stalwarts of the local horticultural society'.[94]

A large factor at play in effecting this post-World War One change was the sudden expansion of the suburbs in the 1920s. New areas were generally subdivided out of farmland. By 1911 the urban population was greater than the rural; generous State Advances Corporation loans allowed increasing numbers of people to purchase their own homes, and by 1916 home ownership rates stood at over 50 per cent.[95] The popular bungalows of the day came in more or less standardised designs, and the majority of sections followed a regular pattern: a low front fence, lawn and a flower garden at the front; utility areas, vegetables, fruit trees and chickens out the back.

Irises in Bert and Annie Wallis's garden, Masterton, 1960s.
Photo courtesy of Gareth Winter

Lyall Sallow and Gillian Jones play golf on the front lawn of a Fendalton house,
Christchurch, c. 1949.
Photo courtesy of Gillian Fox

Working-class gardens: 1940s

Beautification societies increasingly targeted the working classes in the 1940s. In Christchurch, Canterbury Horticultural Society judges inspected the gardens of all state houses annually, whether the tenants liked it or not. The judges were not always impressed; in their report to the society in 1941 they noted, 'on the whole the gardens were a disappointment although in many cases the vegetables were very creditable'.[96]

The beautification of private gardens was still very much an agenda proposed by those in power. In 1946 after the watershed of World War Two, Minister of Agriculture Ben Roberts spoke of what he called the 'mass production of amateur gardeners': 'Landscape gardening, beautification of homes, garden cities, housing projects – those were the modern ways that mass production was taking to elevate the minds and consciousness of the people.'[97] The deputy mayor of Timaru agreed: with the introduction of the 40-hour week, the leisure time people would now enjoy could be spent on improving 'parks, reserves and the citizens' own little plots of gardens'.[98]

The Dunedin garden of J.N. Murdoch, 1958.
Dunedin Horticultural Society records, Hocken Collections – Uare Taoka o Hākena, AG-524-06/001

While many people embraced this ideal, it was often out of a sense of duty, of not letting the side down. But for the most part the beautification agenda was fast becoming accepted as 'right' and 'normal'. The garden plan for the Leeston Methodist parsonage offers a clue to what was expected by 1943. Gone is the emphasis on intensive food production. A concrete path leads along the side of the house to the garage. It is divided by a central strip of grass and flanked on one side by lawn, a rose garden and a bed of hydrangeas and polyanthus, and on the other by lawn and a shrub border. Out the front a low concrete wall shows off the curved shrub and herbaceous border, beyond which a lawn stretches to rosebeds along the front of the house. Copper beech, rhododendrons, forsythia and azaleas are among the many items listed in the plan. The emphasis on flowers, shrubs and lawns in what was presumably a place of high moral gardening standards – especially during wartime – is telling.[99]

The garden societies established throughout Aotearoa early in the colonial period aimed to influence the gardening endeavours of Pākehā colonists. As the societies' emphasis shifted from food to flowers, the limitations of their influence became apparent. Early efforts to get home-owners to beautify their properties gained little traction until the

Not all gardens were abundant or beautiful. Palmyra Street area, Dunedin, 1957.
Dunedin City Council Archive, Photograph 288-2

sudden suburban expansions following World War One. From this point through at least to the end of World War Two, the societies redoubled their efforts to encourage working-class gardeners to cultivate plants for decoration. And while many did create attractive ornamental gardens, the primary focus of the ordinary gardener was still the production of sufficient food to feed the family.

The attention given by garden societies to 'civilising' the working classes may be likened to the notion earlier expressed by Rev H.H. Turton about Māori: in his view, gardening could civilise people. The unruly classes could be made ruly through garden-making. The agenda was effective in establishing a norm for Pākehā home garden management.

CHAPTER 4
Native plants in a fragile paradise: 1914–35

European visitors to Aotearoa New Zealand admired the beauty of native vegetation, and the imperial botanical project – the gathering, identification and sending of plant specimens to the collection at the Royal Botanic Gardens in Kew, London – encouraged early botanists to seek new and interesting specimens of flora. In the second half of the nineteenth century this interest expanded to the general European populace. Motivation for collecting was varied: some tried to earn a living by selling specimens; others 'aspired to join Britain's scientific societies and publish in their journals'.[1] Amateur collecting also became popular in some quarters, aided by cheap guidebooks such as Mrs Stanley Jones' *Handbook to the Ferns of New Zealand* (1861).[2]

A growing appreciation of indigenous flora coincided with 'a gathering sense of nationhood' that resulted from the growing number of 'New Zealand-born Europeans who felt pride in the country of their birth', as Paul Star and Lynne Lochhead have remarked.[3] With this came a desire to explore the wilderness areas of New Zealand. Astute observers in the vanguard of conservation efforts began to comment on the mounting environmental toll of colonialism.

Some gardeners were inspired to recreate native bush or rocky alpine environments in their gardens, and this idea of replicating natural environments coloured garden writing significantly during the inter-war period. As always, however, gardeners were selective about which fashions they chose to adopt.

In his 1844 journal early settler John Williams waxed lyrical about the native trees and ferns in the Bay of Islands; the pōhutukawa, he said, was 'sublimely elegant', a 'most strikingly beautiful tree', 'one of the most heavenly and beautiful sights that the human mind is capable of imagining'.[4] The rātā had 'a superb grandeur ... beyond description', and the kōwhai he considered 'elegant and ornamental'. It formed 'an elegant appendage to the shrubbery' and 'bears transplanting'.[5] Coincidentally, Marianne Williams, wife of missionary Henry Williams, planted a kōwhai with its pendulous golden flowers in her Paihia garden that year.[6]

New Zealand flora captured the interest of the European settlers. In 1842 and 1843 the Wellington Horticultural Society competitions included a prize in the flower section for a collection of native plants.[7] In August 1851 the *Otago Witness* reporter remarked on several native plants in flower, including mānuka, pimelia – 'entitled to be taken notice of ... for its heath-like foliage' – and *Solanum laciniatum* or poroporo.[8] The commercial value of native plants began to be realised too: in 1866 the *Wellington Independent* ran advertisements from W. Kelleway of Pipitea Street in Wellington (formerly the site of a large Māori cultivation), who was keen to exchange exotic tree and shrub seeds for the seeds of native plants.[9]

'The Haszard family and home at Wairoa, c. 1885'. Note the mixture of native plants and exotics in the garden.

Auckland Libraries Heritage Collections, 1677-036

This commercialisation also related to the Victorian fern craze. Frances Loeffler has explored some of the ways in which a market for exotic ferns opened up in Britain later in the nineteenth century. 'Difficult to come by, and impossible to maintain without the expense of a conservatory, or at least a Wardian case, exotic ferns came to function as status symbols and signifiers of personal wealth, an aura of exotic travel and distant lands transforming them from scientific specimens into souvenirs.'[10] The craze, among those who could afford to indulge in it, was fed to a large degree from New Zealand; the significant collection of ferns at the Royal Botanic Gardens, Kew, was certainly boosted by New Zealand specimens.[11]

In Christchurch, 'youthful botanist' Joseph Armstrong's collections of native flora were described in the newspaper and excited interest at horticultural exhibitions. In 1864 he discovered moss and fern species overlooked by the director of Kew, British botanist and explorer Joseph Hooker, and earned a guinea and a half for his dried specimens.[12] In 1872 Armstrong displayed 180 ferns and fine-foliage plants at the horticultural show, about which there was a lengthy write-up in the *Press*, and

32 Edinburgh Street, Spreydon, Christchurch: a garden with a generous planting of natives.

Hean Collection, Canterbury Historical Association Collection, Canterbury Museum, 2000.198.511, CHAC511

he was still exhibiting specimens 10 years later.[13] That Armstrong's ferns gathered such attention at these shows is an indicator of who was most interested in them – in other words, those who attended such shows: likely the middle and upper classes.

There was a growing interest in alpine plants, too. At the 1884 horticultural show Armstrong 'exhibited the peculiar Alpine plant known as the vegetable sheep' (*Haastia pulvinaris*), and Christchurch seed merchants Adams & Sons – who dealt in alpine plants such as aciphyllas and alpine veronicas (or hebes) – had a number of alpines on display.[14] By 1894 this company was promoting itself as the 'Establishment for New and Rare Plants', including native species.[15] New Zealand's first comprehensive gardening manual, Michael Murphy's 1895 *Handbook of Gardening for New Zealand*, incorporated a section on native plants including alpines for the rock garden, information for which was supplied by Adams & Sons.[16]

Imperial gardening interests

The Royal Botanic Gardens, Kew, in London, was a centre for scientific research and the international exchange of plant matter, and contained specimens from all over the world. The director – William Hooker from 1841–65, followed by his son Joseph from 1865–85 – took great interest in New Zealand discoveries and welcomed specimens and seed for the collection.

The Canterbury Acclimatization Society in 1867 sent live native birds, packets of native seeds and a case of native plants to London. Kew Gardens also supplied plants to New Zealand: scientist and explorer Julius von Haast wrote numerous letters to Kew in the 1860s to request specific European plants for the new government domain in Christchurch and offered fresh discoveries of New Zealand native plants in return. Haast brought the botanising work of John Armstrong, father of the aforementioned Joseph and later the government gardener at the Christchurch domain, to the attention of Joseph Hooker: 'I send … a parcel of seeds, all collected by J.F. Armstrong … I hope they will be of value to you and germinate.'[17] Two years later John Armstrong was tasked with filling Wardian cases with specimens and sending them to Hooker at Kew. Haast reported, 'I have procured for him [Armstrong] leave of absence … to go into the Alps in the proper season, to collect the live plants you want.'[18] Armstrong obviously had a good eye for new plants: in 1870 Haast wrote to Hooker, 'Armstrong ... tells me he has got some more novelties.'[19]

In 1886 Adams & Sons wrote to botanist William Thiselton-Dyer, Joseph Hooker's son-in-law and his successor as director at Kew (1885–1905), asking him to retain all plants sent by Adams 'likely to be of commercial importance'. These included the large-leafed forget-me-not *Myosotidium hortensium* and alpine coprosmas that would thrive in colder parts of England.[20] Thiselton-Dyer's response is not recorded but was likely favourable: he had overseen the construction of Kew's rock garden in 1881 and the adjacent alpine house in 1887 – both important markers of the development of imperial gardening interests.

People continued to send plant material from New Zealand to Kew into the 1920s: Sir Frederick Chapman (who had discovered *Leptospermum scoparium* var. *Chapmani*, the red mānuka) posted seed of 'a very attractive Carmichaelia' that he found near Franz Josef glacier to Arthur Hill, now Kew's curator, in 1927.[21] Later that year Hill began communicating with Leonard Cockayne, whose reputation as a skilled botanist with an interest in hybridisation was well known both in New Zealand and abroad. Cockayne was a stimulating correspondent. Their discussions touched on celmisias and olearias that seemed to be hybridising in response to the opening up of shrubland. Hill was keen to visit New Zealand: '[T]here is nothing I should enjoy more than seeing the New Zealand botanists and something of the vegetation of the country.'[22] Cockayne was delighted: he offered to accompany Hill from Christchurch to Franz Josef glacier, and suggested they walk from Arthur's Pass railway station to Ōtira station via Arthur's Pass and the Ōtira Gorge so that Hill might see the mountain vegetation for himself.[23] Hill was excited by what he saw and claimed to be 'more impressed with the West Coast than anywhere'.[24] He received a warm civic reception in Christchurch and was much touted by the Canterbury Horticultural Society.

Conservation concerns 1900–35

Jonathan West has chronicled the changing environment of Otago Peninsula, writing that by 1900 the settlers 'had entirely altered almost every aspect of their surroundings':

> Almost all of the native species ... had been eradicated in most
> areas. The indigenous trees, shrubs, ferns and even mosses and
> lichens were largely gone ... Most of the native birds were gone ...
> Even the insects were almost wholly different. From the soil up, the
> land had been remade in the settlers' image.[25]

Concerns about the negative environmental impacts of colonial settlement were increasingly voiced through the 1890s.[26] Attention was shifting from an urgency to clear the land of trees to make way for farms and settlements, to anxiety, felt throughout the British empire, about the relationship between deforestation and the loss of flora and fauna, to growing consternation in New Zealand regarding the loss of bird and insect life, the climatic impact of deforestation and increasing problems of sheet erosion.[27]

These concerns were noted in Canterbury. The Christchurch Beautifying Association, of which Cockayne was a founding member, busied itself in 1900 with the preservation of 28 acres (11ha) of native forest named Kennedy's Bush, on the range between Christchurch and Lyttelton known as the Port Hills, to 'provide a haven of refuge for our New Zealand birds now rapidly becoming things of the past, and where also might be conserved and gathered together all kinds of native trees and plants ...'[28] In 1906 the association became involved with the public ownership of Riccarton Bush, a remnant stand of kahikatea floodplain forest, and in 1908 planted a number of cabbage trees (tī kōuka) 'near the Convalescent Home, on the Cashmere Hills'.[29] Christchurch city councillor and Member of Parliament Harry Ell, a champion of Kennedy's Bush, also turned his attention to conservation issues around the Port Hills at this time.[30]

Concern about the state of New Zealand's forests led to the establishment in 1913 of a Royal Commission on Forestry, of which Cockayne was a member. The commission was 'charged with the examination of existing indigenous forest land to determine which of it ought to be retained for purposes of soil protection, water conservation, and scenic purposes'.[31] According to the commission's final report, indiscriminate deforestation needed to be reined in:

> ... few countries in the world are ... more in need of an adequate
> forest covering on their high lands than is New Zealand. The lofty
> mountain-ranges which traverse both Islands and the excessively

broken nature of the land in many places, together with an average high rainfall, lead to the presence of innumerable streams, and offer ideal conditions for denudation. In other words, the mountains and hills of New Zealand would, if not forest-clad, be a constant source of danger to the farmlands on which the prosperity of the Dominion so greatly depends.[32]

In 1923 the Native Bird Protection Society (later renamed the Royal Forest and Bird Protection Society, or Forest & Bird for short) was established, in an effort to avert the 'dire calamities' deemed likely to result from continued deforestation.[33] Michael Roche has pointed to Forest & Bird's preoccupation with soil erosion throughout the 1930s, but this focus was already evolving in the 1920s.[34] In 1924, in the first volume of the society's publication *Forest & Bird*, E.V. Sanderson wrote: 'Civilization and progress would end with the depletion of the soil and the great increase of insect life. This is especially applicable to mountainous countries like New Zealand, where erosion is severe.'[35]

The perception of New Zealand as a fragile environment began to be reflected in gardening literature in the 1920s. The *City Beautiful*, a monthly journal published by Canterbury Horticultural Society, made reference to farming practices, such as burning, that were gradually destroying areas where interesting native species could be found in the wild. In a 1928 article on celmisias the author commented that on Titan Ridge in Otago's Garvie Mountains one could walk through *Celmisia lyallii* for 'a couple of miles':

… the plants are growing so closely together, as almost to touch each other. On the opposite side of the Ridge, C. Coriacea – probably the king of Celmisias – must cover hundreds of acres, making the mountain slope one unbroken sheet of snowy whiteness! The reason why there is such abundance here is that the owner of this 'Run' is not an advocate of 'burning' as a factor in the improvement of sheep pasture. Would that all runholders were imbued with the same principles! It would remove the menace that before many years are passed nowhere but in the inaccessible places will any evidence of our beautiful native flora be seen.[36]

Another article, this one on *Clematis indivisa*, noted that where the white-flowered climber had once been common on the Port Hills,

Front cover of The City Beautiful, *November 1929. This Central Otago scene shows the influence 'back to nature' thinking was having on the Canterbury Horticultural Society at the time.*

Canterbury Horticultural Society

'Twenty-five years of tussock burning ... have robbed us of the bush and of its clematis.'[37]

Tourists and plant collectors were also causing degradation. G.R. Butler of Arthur's Pass wrote to the *City Beautiful* in April 1928 to express concern about the impact of tourism, a growing issue in that area since the opening in 1923 of the rail tunnel through the Southern Alps at Ōtira. According to Butler, the influx of 20,000 visitors annually was subjecting the area to 'spoliation and destruction': '[It] is not too much to say that the native plant life and beauty of the most accessible parts are in grave danger of being irreparably damaged.'

Some parts most easy of access have already been practically denuded
of plant life. A particular example of this can be seen on a spur just
off the road on the Otira side of the White bridge which was, a few
years ago, a veritable garden of Celmisia Coriacea. Now all that is left
is but a few of the poorer plants … Protection and development of the
Arthur's Pass National Park is a matter of deep concern to Canterbury, and particularly to the city of Christchurch.[38]

The issue also contained a full report regarding the establishment of a board of control for the soon-to-be-created Arthur's Pass
National Park.[39]

Rock and alpine gardens 1895–1940

We might assume that the period of most vigorous promotion of native
plants would have coincided with the so-called 'Māoriland movement',
roughly 1890–1914. This was essentially a translation, to the New Zealand context, of the British Arts and Crafts movement, with its interest
in natural forms such as plants and fine arts and crafts. It is epitomised
in Aotearoa by the Pākehā appropriation of Māori motifs and themes,
and is also reflected in literature of the period.[40] In fact, Pākehā gardeners showed more interest in native plants in the years immediately after
Māoriland. The timing coincided with the rise in national consciousness after the watershed of World War One, and the growth of a New
Zealand-born Pākehā population. However, the interest in New Zealand
native plants appears to have been another short-lived, middle-class
Pākehā garden fashion, as explored below.

Pākehā suburban gardens were undergoing change. The almost
exclusive focus on food production in fertile New Zealand soils was
broadening to include beautification projects. Rock gardening, a tradition well known to the British, was one feature in which New Zealand
plants were easily incorporated, and in this a growing awareness of the
fragility of the New Zealand environment – particularly of alpine areas
– found new representation in some home gardens. In his first edition
of *Manual of Gardening in New Zealand* (1914), David Tannock described
the virtues of developing a rockery: '[If] it is desired to hide an ugly or
undesirable feature either in your own or your neighbour's garden, there
is nothing so satisfactory as a miniature hill or a range of mountains.'

The design of a rock garden will depend on its extent. If small, it is better to reproduce one mountain peak, sloping up irregularly; if space will permit two peaks can be formed with a ravine or shingle slip between, and this can be extended until a whole mountain range is reproduced. One can learn a lot from nature both in design and placing the stones, so that before commencing operations it is well to study a mountain range and to form a mental picture of what you mean to accomplish.[41]

In his *Handbook for Gardening* Michael Murphy promoted the use of native alpine plants, which he considered an under-appreciated resource, and congratulated Adams & Sons for making these more widely available, saying 'a collection of alpines is a never-ending source of interest'. 'Few persons, except those who have visited the Alpine regions of New Zealand, have any idea of the beauty of the flora.' The actual gathering of plants in the wild was, he said, 'a great source of pleasure and relaxation'.[42] Naturally, the rock garden was the perfect place to display these.

A. Wilson promoted a similar message in Dunedin in 1901: 'One of the greatest natural glories of this island is its noble alpine chain; and its alpine flora is exceedingly interesting to the mere gardener. It is quite possible to stock an alpine garden of considerable dimensions from our own alpines alone.[43] James Speden, writing for the *City Beautiful*, suggested, 'To see these plants in their natural habitat on the tops of those rocky ridges, imbues us with the desire for a similar rockery in our gardens, a transfer of our mountain ideals.[44]

Those considering developing a rock garden were encouraged to consider more than just native species. An article in the *Star* in 1906 promoted pinks from Austria and Transylvania, and in 1910 the *Press* recommended a number of Chinese plants.[45] The Nelson-based New Zealand Alpine and Rock Garden Society, whose members were involved in plant-finding expeditions in the 1920s, was interested in obtaining exotic plants from Kew, and in 1929 and 1930 various authors wrote affectionately of campanulas from southern and eastern Europe as their preferred rockery plants.[46]

From the 1920s a clear advocacy emerged for the use of natives in the rockery. Leonard Cockayne wrote enthusiastically about veronicas and a host of other native plants in his 1923 seminal work *The Cultivation of*

New Zealand Plants.[47] A reviewer in *New Zealand Truth* expressed astonishment at Cockayne's suggestions: 'We are so accustomed to think of English flowers … that it comes as a surprise when we think of our almost entirely neglected New Zealand plants for this purpose.'[48] The *Truth* writer was all in favour, however, and the following month reported that the 'cult of the native flora is gathering force – and certainly not before its time': New Zealanders, the reporter now felt, should make more of an effort to include natives in their gardens and parks.[49] A similar theme was pursued in a 1927 lecture to the Otago Women's Gardening Circle: the speaker hoped the cultivation of native alpine plants would 'excite the interest of many, so that [alpines] would become familiar to lovers of native plant life'.[50]

The *City Beautiful* provides evidence of an increasing Cantabrian fascination with native plants, and with alpines in particular. Almost every issue of this journal from 1928–30 included an article on either rock gardens or alpine plants for rock gardens – usually natives.[51] Other articles championed particular native species, such as the adaptable celmisia or alpine daisy.[52] When newspaper proprietor Sir George Fenwick was asked to provide a guest editorial for a special 'rose issue' of the publication in 1928, he wrote that he would rather give 'some references to the success with which our native Alpine plants are grown in Dunedin'. He hoped nurseries would 'spread their efforts to the systematic growing of the best of our Alpines. They are full of interest for lovers of our native vegetation, and ought to be found more freely in the private gardens of Christchurch.'[53] Ivory Brothers' catalogue for 1927–28 included a number of native plants, and J.M. Baxter's native plant catalogue for 1928–29 listed 62 veronicas.[54] For a rock garden, suggested Christchurch horticulturist Morris Barnett, 'one need not look beyond our own native plants for suitable subjects. The whipcord veronicas, dwarf Helichrysums, Pimeleas, and last, but by no means least, the New Zealand pigmy pines … provide excellent specimens for this purpose.'[55]

Despite all this promotion, entries in the first horticultural society rock-garden competitions in Christchurch were not high: 11 in 1932, 12 in 1933 and 1934, 10 in the next two years and just eight in 1937. Thereafter the record is imprecise until 1945, when there were just three entries, and by 1950 rock gardens no longer figured in the programme.[56] H. Vincent, winner from 1932–34, wrote several articles on rock gardens in the

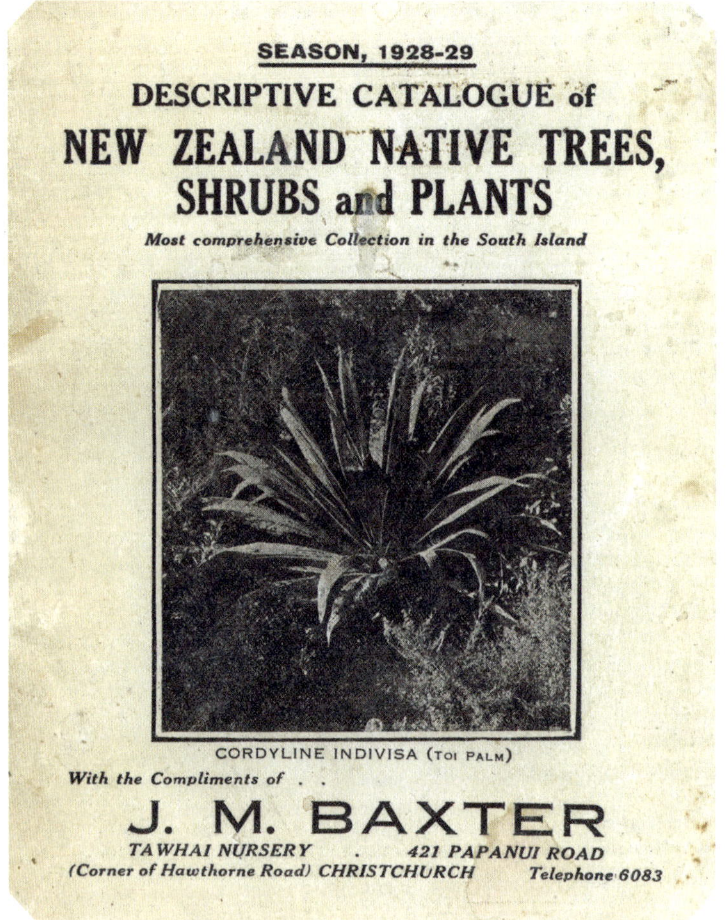

J.M. Baxter's native plant catalogue, 1928.

City Beautiful, one of which provided rich detail of a Cashmere garden that contained '400 to 500 varieties of alpine plants', including 'many species drawn from mountains from all quarters of the globe':

> There are plants from the Himalayas, the Swiss Alps, the Balkans, and the Andes. Siberia, Russia, Greece, Italy and Spain have delivered the treasures of their mountains. Even the lonely Falkland Islands are represented. Japan and China have supplied primulas for the beds at the feet of the rocks. And celmisias and other herbage from the New Zealand Southern Alps flourish among their brethren from overseas, their thick foliage contrasting against the shorter-spiked saxifrages or nestling campanulas and gentians.[57]

According to Vincent, a rockery should capture 'nature in her wild moods ... loveliness, maybe, in some boulder-strewn hillside where the ground is carpeted with close-knit verdure', with 'crannies of some weather-beaten rock ... The aim is informality, whether of rock or of general outline.'[58] He encouraged local rock gardeners to become familiar with European species, such as *Veronica teucrium trehane*, whose 'foliage is bright golden, so that at a distance one is deceived into believing that the gold is that of blooms ... [At] flowering time ... spikes of sapphire blue appear, giving a contrast almost ethereal against the yellow leaves.'[59]

Vincent also showed appreciation for native alpine flora and assured readers that he could discuss them 'for hours'. In the final article of his series he praised New Zealand alpines: 'veronicas (or hebes) should be a strong feature of every rock garden ... For foliage effect, have several "whip-cords".'[60]

The dual fascination with exotic and native alpines is reflected in the judges' comments for the 1932 competition:

Each rock garden ... should have a character of its own. Standing in the middle of one of the gardens inspected, one might well imagine oneself to be in the heart of New Zealand mountain country, with rugged tussocks growing over the rocks in the more exposed places, dainty flowering plants in the sheltered and sun-warmed hollows, and delicate ferns growing in profusion in the lower and more moist sections of the garden. In others the brilliance of flower and foliage gave a strongly tropical touch to the surroundings.[61]

According to the judges, only true alpines should be present in rock gardens: 'The aim should be to grow as comprehensive a collection representing the various groups of beautiful alpines from the various alpine regions of the earth, not forgetting our own New Zealand alpine flora, as it is possible to do with the amount of space at one's disposal.'[62] While the presence of natives was 'gratifying', exotics would provide 'the colour that New Zealand plants lack'.[63] Five years later little had changed. The 1937 judges praised R. Sladen's garden for the exotic rarities it housed, and for his 'good collection of New Zealand alpines'.[64]

The idea of recreating an alpine area in a suburban garden was carried back to the mountains by one keen young horticulturalist. Writer

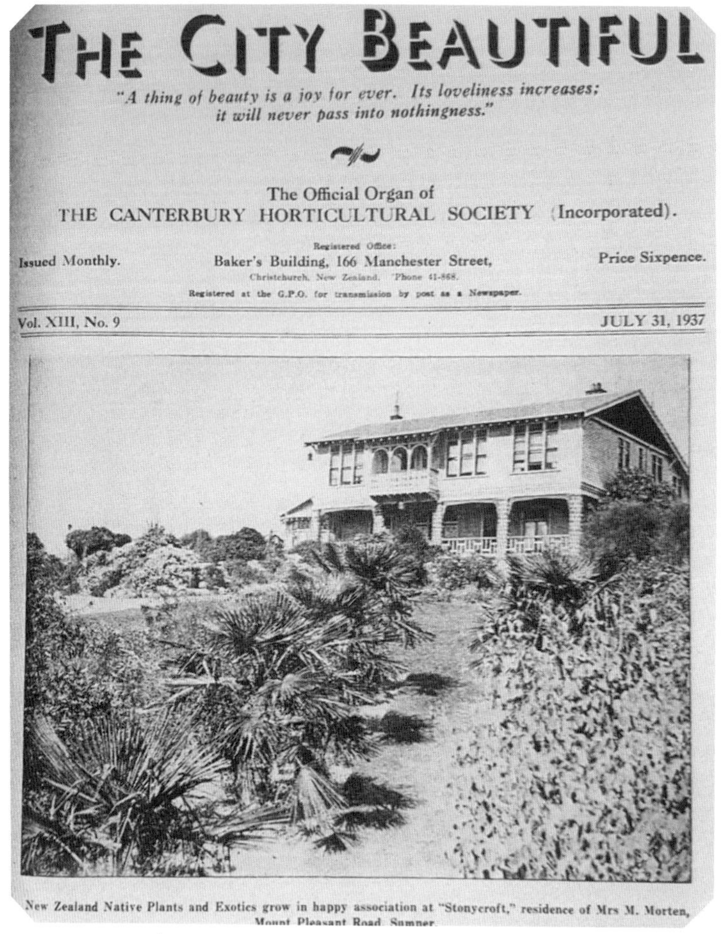

The front cover of The City Beautiful, July 1937, showcases a hill garden
with the desired mixture of New Zealand native plants and exotics.
Canterbury Horticultural Society

Jean Lawrence described a garden developed in the harsh environment
of Ōtira, just beyond Arthur's Pass in the Southern Alps. Rocks and
boulders pulled from a levelled area became rockeries bordering gravel
paths; a natural bank was home to ferns and mountain beeches, and 'naturally regenerating gayas' were allowed to stay. The gardener included
mountain lilies and mountain daisies – plants Lawrence claimed were
simpler to source then, as 'in those early days [prior to the establishment of the national park] there was no prohibition on the taking of
native plants'. Hebes, celmisias and gaultherias were planted in a central

garden, and between them 'ordinary English garden flowers began to grow', including Russell lupins (which ultimately became a pest in the area). At 600 metres altitude this flower garden was rightly considered 'an audacious project'.[65]

Plant collectors in the wild, 1915–35

Despite concerns about growing environmental degradation, the desire for native specimens took gardeners into the wild where they helped themselves liberally. In 1915 and twice in 1917, a Dunedin-based women's gardening circle organised a fern-collecting expedition to St Leonards (on Otago Harbour) where the Cook sisters showed members how to find the best specimens.[66] The 1922 obituary for Margaret Johnston of Ōtautau, Southland, praised her as 'a keen naturalist' who took 'great interest in native plants and shrubs. During her rambles she made many collections and successfully transplanted many of them round her home.'[67] She was but one of many doing the same.

Avid Dunedin gardener Alexander Paterson developed a collection of New Zealand native plants on his harsh south-facing slope in the mid-1920s. He was especially keen to obtain 'moderate supplies of proper rockery plants that are not readily obtainable in trade channels', and offered to pay the Nelson Rock Garden Society to go on plant collecting jaunts to find particular ferns that did not grow in his region.[68] He had, he said, plenty of tōtara and a 'full stock' of rimu, but was happy to receive celmisias, and hoped the group would collect some young nīkau plants for him 'from the real Alps'.[69] 'How I envy you the collecting outings you mention,' he wrote wistfully.[70]

The gradual expansion of rail and road corridors brought mountain and bush areas within easier reach, and the enthusiasm for planting natives in suburban gardens began to have a discernible impact in some places.[71] In 1928 James Speden warned gardeners against the indiscriminate uprooting of plants in the Arthur's Pass area, saying it would ultimately ruin 'one of the most beautiful and interesting walks in New Zealand'.[72] A. Tyndall's article, titled 'The Destruction of Our Native Flora', had a similar message, and a public plea was made in the *Press* in 1935 for collectors to gather seeds, not plants, from the wild.[73]

June Stewart's parents had a cottage at Arthur's Pass in the 1930s: 'Mother was keen on tramping and climbing, so she introduced us to the mountains,' June recalled. Her parents established a native section in their garden on the original Deans estate in Christchurch, which contained matāi, kahikatea, rimu, kōwhai and other trees. The plants were largely sourced from the very environments Speden and others were writing about. 'We felt that we would have our own little bit of bush ... Mother brought back the rimu from the West Coast ... We brought a clump of stuff we found at Ōkārito, which turned out to have the kahikatea and other nice natives in it.' They also established a rock garden, which June referred to as a place of 'associations': each plant had a story and bore a reminder of the area from which it was sourced.[74]

This blithe plant collecting was in some ways the downside of the 'back to nature' movement of the inter-war years, when the popularity of tramping, mountain climbing, camping and exploring the great outdoors soared. New Zealanders appreciated the beauty of the bush and took to the hills to enjoy it in increasing numbers. In a 1932 article about ferns and ferneries, horticulturalist James McPherson, who had been trained by David Tannock, had spent time at Kew and been curator at Invercargill Botanic Gardens since 1926, wrote, 'there is nothing that stirs or impresses me more than sitting on an old rotted log in parts of our New Zealand bush, whether lowland or highland, and drink deep of the beauties of the ferns around me ...'[75]

He was not alone in his sentiments. Getting 'back to nature' was a healthy antidote to urban living, but as city-dwellers took to the countryside many were confronted by the degradation resulting from timber milling and the clearance of land for farming. Kathleen Guy recalled her mother's passion for the bush: 'I can remember going with her to the West Coast when I was little. On a walking trip ... from Murchison to Greymouth ... my mother burst into tears because the forest that she had seen before I was born was gone.' The destruction in the country's hinterlands motivated Kathleen's father to plant native species around their own home: 'My father was passionate about birds. He was an ornithologist, as a hobby, and a lot of the trees were planted [at the family home] to feed the birds.' Kathleen remained on the property after her parents died, and continued to develop the bush slopes.[76]

Christchurch hill-garden competitions

Some of the worst sources available to historians of 'ordinary' gardening are the reports on prize-winning gardens, because these gardens simply aren't ordinary. However, where the gardens themselves diverge from 'best practice', as enunciated by the purveyors of good gardening, such studies can be illuminating.

Farms in the Canterbury Port Hills were gradually subdivided during the 1920s and the number of north-facing hillside residential properties increased. Ecological conditions on the dry sunny slopes were completely different to 'the flat', and many, particularly those near the sea at Sumner, were largely frost-free.

Some of these hill gardens were described in the *City Beautiful*, which at this time was strongly advocating native gardens. Surprisingly, however, one of the main themes of this material is the emphasis placed on botanical eclecticism. The globalised garden was 'in'. Jack Humm, in his 1930 article, commented on the conditions in the hill suburbs: 'The climate is very favourable to sub-tropical trees and shrubs, and many rare and beautiful subjects grow and flower most luxuriantly there.' Australian and Peruvian plants thrived, and one garden had plants 'drawn from the four corners of the earth … growing together in perfect harmony'.[77] Another contained what he considered the finest pūriri in Canterbury, together with tree ferns, lemon trees, a pōhutukawa, a splendid *Quercus ilex* (holm or evergreen oak) and specimens from South Africa and Mexico.[78] One owner could 'truly claim to be a citizen of the world, for as she walks round her truly delightful garden, she can see almost every country represented'.

The Canterbury Horticultural Society initiated a hill-garden competition in 1929, and the Sumner and Redcliffs Beautifying Association ran a separate competition, also from 1929, which featured both hillside and flat gardens. C.H. Reece wrote in the *City Beautiful* that W.J. Sim's Clifton garden featured Australian 'Grevilleas, Callistemons, Acacias, Eutaxias, and numerous others of her noble band', along with South African proteas, Californian prickly pear and 'several others of the cactus tribe'.[79] In Cashmere, Lyndsay Russell's winning Alfred Buxton garden contained an impressive rockery planted with many low-growing azaleas, and John Macmillan Brown wrote at length about the exotics

A small decorative rockery in Dunedin, 1928.
Dunedin City Council Archive, 293-5

on his property, especially those from Australia, South Africa, Japan, China, North and South America and the Pacific.[80]

Native plants were accorded a special importance: they were graceful, charming, delicate, and their indigeneity marked them as worthy of attention. One report noted, 'Most of the hill gardens … fully exploited the usefulness and beauty of native trees and shrubs, and few … were without the popular kowhai.'[81] Sim's Clifton garden featured the New Zealand lilac *Heliohebe hulkeana* along with '[m]any native plants', and T. Smart's winning five-acre (2ha) Cashmere property was noted as a fine example of how gardens could be made to replicate native forest.[82] As well as enumerating his collection of exotics, Macmillan Brown described his many native plants, including those that grew over his rockeries.[83]

Kate Jordan noted a similar co-mingling of exotic and native species some years later in William Ball Dutch's Wellington garden, developed from 1953. There she observed what she called an 'internationalist–nationalist' spirit at play: as New Zealanders became better acquainted with the 'diversity of the world's plants', she felt they had also discovered 'the uniqueness of their own plants and became proud of them'.[84]

Entries for the Christchurch hill-garden competition declined from 1946.[85] In 1947 the *City Beautiful* reported: 'For some strange reason the number of entrants in the Hill Garden Competitions judged on November 5 was very small. The season has been good and interest in this type of garden is well maintained, so that it is difficult to find a good reason for the paucity of entries.'[86] By 1951 things were dire: 'The Society's Committee is anxious to see more entries, in order that the Hill Gardens Competition may be continued. Without more competition the Society may be compelled, reluctantly enough, to abandon it.'[87] And so it was.

A native plant craze?

Prizewinners were exemplars of fashion, and few hill gardeners could claim to have gone overboard on native plants. Surprisingly, the same is true of those gardens that won the Canterbury Horticultural Society's 'native garden' competitions. Te Wharekoa, a property in Banks Avenue in the Christchurch suburb of Dallington, regularly won first place, even though the 1.2-hectare property was not an exclusively 'native garden'. As in many gardens, the incorporation of elements such as alpine and native bush zones was complemented by plantings of exotic species. Winifred Chapman described Te Wharekoa in rich detail in 1930: 'Plantations of tree and shrubs occupy, as it were, the wings, whilst gay gardens of summer flowers are disposed about the chaste white marble sundial on the higher levels of the lawn.' The curved gravel driveway took visitors between 'a grand old monkey-puzzle, junipers, copper beeches and spiraeas'. A sloped area featured a fountain and crescent-shaped beds of azaleas and rhododendrons. 'Beds planted with rose trees, pink antirrhinums, and catmint make a brilliant display. The bronze of marigolds lightens the ground beneath the graceful weeping trees which adorn the lawn ...'[88] Not many natives there! However, native plants were 'an especially attractive and decorative feature of the garden': ribbonwoods, olearias, veronicas, senecios, kōwhai, rimu and lancewoods, 'whose tall slender trunks bear tufts of dark spear-like foliage at the top', were all present. 'One well-grown rimu, with its fragile drooping branchlets of palest green just tinged with brown is worth going far to see, as is also the unexpected beauty of the bracken in the native rock garden,' wrote Chapman.[89]

As with the hill and rock-garden competitions, the lacklustre uptake of this competition must leave us wondering just how popular native plants (and garden competitions) were during the interwar period. Many people were buying natives – gardening catalogues provide ample evidence of this – and those who weren't purchasing them were uprooting them from the countryside. In a survey of property advertisements for Christchurch, however, native plants, which began appearing from the 1930s and were slightly more prominent in the 1950s, barely register as a significant element. They figure in the hill suburbs, in affluent Fendalton, and in Papanui and Riccarton.

Oral histories from residents in these areas indicate that gardeners had a broad taste in plants. Audrey Potter was born in 1933 and lived in front of Riccarton Bush; next door was a former Deans family gardener who was 'interested in native plants' and once gave her a tour of Deans Bush.[90] According to Audrey, natives didn't feature in her family's garden. Judith Todd, in nearby Harakeke Street from 1940, felt that native plants would have been out of place in what she called her 'English garden'.[91] Sanger Holmes, who kept native plants tidily trimmed along the Avon River border of his property, said his main passion was dahlias.[92] Greystones, a Fendalton property built in 1926, featured a large fernery even in the late 1940s, but the copper beech, weeping elms, azaleas, hydrangeas and expansive lawn were the most striking features of this garden.[93] Alison Helm, also in Fendalton, tried to make a rock garden in the 1950s but without success. The specialised plants were 'so tiny' she was 'apt to lose them'. She did not specifically grow natives either.[94]

In the hill suburb of Clifton, where volcanic rock is plentiful, gardens were noted for their informality, native plantings and rockeries, but here too residents exhibited a wide taste in plants. Fern Every's special passion, in her 1938 garden, was for sunflowers and corn, which reminded her of childhood days in Ohio and China.[95] Hugh England, a resident of Kinsey Terrace from 1928 on a property originally owned by Christchurch florist Shillito, still maintained a fernery supplying greenery to the local flower market. The Englands grew tamarillos, figs, tobacco, hazelnuts, almonds, walnuts, varieties of plums, cherries, peaches, citrus, blackberries, gooseberries, blueberries and many others, and especially plants of South African origin donated by Joseph Kinsey, who lived next door.[96] Kinsey was passionate about native plants and grew them in his rockeries, but also incorporated plants from all over the world.

The Guy family, who bought the Englands' property in 1945, ultimately replanted it in native bush: 'We were brought up in the mountains, and I like the way they make us feel small,' Kathleen Guy mused. Reflecting on her untidy bush garden, she commented: 'This place tells you … like the mountains do, that we are very transitory, that we are just one

An impressive mix of natives and exotic plants in Joseph Kinsey's rock garden on Clifton Hill, Christchurch, c. 1900.

J.J. Kinsey, Canterbury Museum, 1951.35.186

of the things that live here, and that other things go on beyond.'[97] Yet Kathleen's father was committed to his fruit trees and vegetable garden, while her mother maintained a large rose bed, complemented by plantings of daisies and Christmas lilies.[98]

Prue Lovell-Smith's mother lived in Cashmere from 1936. She loved rock gardens and grew both olearias and koromiko, but focused on her 'beautiful flower garden'.[99] Grace and Guy Butler (champion of Arthur's Pass protection) lived in Cashmere in 1923, and Grace used to take her granddaughter Jennifer Barrer exploring in the bush. Jennifer's father Brian was an avid mountaineer, and her mother Margaret was married with *Ranunculus lyallii* in her hair. At the Cashmere Garden Club, which Margaret established in 1940, she played films of the Southern Alps and organised Morris Barnett to come and speak about New Zealand natives. These of all people might have been expected to have a strong focus on native plantings, but in fact Brian's passion was gladioli, and Margaret sought to emulate the eighteenth-century landscape gardener 'Capability' Brown, renowned for his sweeping lawns, clumps of trees and informal ponds or lakes.[100]

Margaret Barrer's lily pond – 'emulating landscape architect Capability Brown'. Cashmere Hill, Christchurch, 1940s.

Photo courtesy of Jennifer Barrer

Early twentieth-century Māori gardens

Literature on gardening with native plants from this period rarely mentions Māori, and then usually only in the past tense: Māori *used* to live in harmony with nature; colonisation *had wreaked* unimaginable damage in the forests, therefore Pākehā must rectify this in their gardening. Māori were still mostly living in rural areas, and some communities lay beyond the reach of civil authorities and organisations that were trying to preserve native flora or build a sense of nationhood.

There is little evidence to suggest that Māori were actively encouraging native plants around their dwellings. Florence Harsant (neé Woodhead) recorded her memories of gardens in the Māori village of Waitahanui on the shores of Lake Taupō, where she lived from 1905: 'The [Māori] cultivations were used for growing potatoes, kumara and sweet corn and hue [gourd], a marrow-like plant which they would shape as it grew & use as a water carrier … For greens they used watercress and puha [sowthistle].'[101] With the help of some Māori men Florence's father cleared the land around their house and brought in soil from the swamp.

'Father planted vegetables,' she recalled, 'and Mother and I, flower seeds.' Unlike the Woodheads' mixed garden, the pā cultivations were entirely devoted to vegetables, many grown from seed supplied by the government. Māori did not always use vegetables in the European way, however: 'We went down to the pa one day to see the vegetable garden, and found carrots, turnips and parsnips all strung up to dry.'[102] The leaves of beetroot were used and the roots thrown away. 'Lettuce and cabbage … were used in the same manner as "puha".'[103]

Some evidence exists of efforts to alter the character of Māori communities through gardening, for example in the work of the Waikato Māori Committee, which was established to address issues affecting local iwi. The committee papers include newspaper articles on gardening practices from lawn-making to rose growing and spraying, but nothing about propagating native plants. A 1919 article stressed that 'The work of making a new lawn being of a permanent character, no labour or expense should be spared in having it well done at first.'[104] It seems unlikely that a perfect lawn would have been considered a priority for rural Waikato Māori, who had been uniquely subjected to conscription and harsh treatment during World War One and then faced the devastation of the 1918 influenza epidemic.[105]

The story of the Tomoana whānau provides some insight into gardening practices of Māori who still had access to land. Paraire Tomoana (1874/5–1946) and his wife Kuini Rīpeka Ryland planted their garden at Waipatu, near Hastings, according to ancestral knowledge, traditional techniques and the phases of the moon. Their son Tana was born in 1911 and grew up working alongside them, learning such things as how to get the best mix of sand, rotted plant matter and hay to grow the kūmara tipu. In the 1930s they diversified their crops to include sweetcorn, melons, beans, tomatoes and other vegetables. Tana married Elva, of Ngāti Maniapoto (Tainui) descent. Elva was closely involved in the vegetable garden, but she also grew a mass of hollyhocks, geraniums, roses and hydrangeas.[106] While growing native plants did not feature here, in this description we can see a striking similarity between Pākehā and Māori approaches to gardening.

For Māori without land the situation was quite different. Photographs from the late 1930s, when the government began to construct new rural dwellings for Māori 'indigents', illustrate this. Few people in these communities would have had the means or desire to plant fashionable shrubs – native or otherwise.

An antipodean vision

Horticultural exhibitions and garden competitions encouraged suburban beautification and diversity of planting. The competitions – essentially civic rituals – helped to instil a changing perspective into an established pattern and, in the process, transformed that pattern. Pākehā may have domesticated native plants, but it is difficult to detect in this effort an abiding devotion to any fixed British idea. Particularly in the 1920s and 1930s, the impassioned environmental philosophies attached to growing native plants suggest a different motivation. In effect, a new form of garden emerged in which the New Zealand environment could speak. Mimetic gardening – in this case the reproduction of a natural form in the garden – created antipodean versions of British garden forms, intensely aware of the immediate environment but not exactly hybrid. It was perhaps part of resolving the uncomfortable contradiction of 'belonging and not-belonging' that Ngaio Marsh felt when she ventured from Cashmere into the Southern Alps during World War One, and which Ursula Bethell felt when musing on her primroses: 'The sight of you here under the apple-tree has too sweet a sting. / So like, so unlike the sight of you in an English garden in spring.'[107]

People continued to emphasise the importance of conserving native flora and fauna. A 1936 article in *Forest & Bird* urged New Zealanders to plant vegetation that would 'attract into gardens colonies of songsters and feathered friends'.[108] The journal also ran articles on native and bird-friendly gardens, such as a 2000m^2 property at Paekākāriki, near Wellington, belonging to Captain Sanderson, a key player in Forest & Bird. Over 10 years he had transformed a 'dreary stretch of sanddunes' into 'a delightful young forest, where one may enjoy that bush scent which warms the heart of nature-lovers'.[109] Baxter's Tawhai Nursery was also singled out for attention. Situated on the slopes of Mount Pleasant in Christchurch, it covered 1.2 hectares: 'Mr Baxter's garden is an excellent example of the results of planting "bird food" and supplying fresh drinking water. Birds are with him all the year round.'[110] They were early efforts to recover what one writer in 1946 called the 'lost avian paradise'.[111] Others moved to conserve areas of bush and develop new native plantings. At Waihi Bush in 1947, Ethel Musgrave resolved to make the 14 hectares of bush on her farm – the only remaining sizeable stand of relict forest on the Canterbury Plains – into a reserve. It would be placed under a QEII National Trust covenant in 1981.[112] And as post-World War Two development further threatened bird habitats, Forest & Bird continued to urge readers to plant with birds in mind: 'almost every suburban or provincial garden can be made to figure in the bird-attracting campaign'.[113]

The debate about soil erosion took a new turn in 1937 with the creation of the Native Plant Preservation Society, which made a bold statement in its first annual report: 'The present generation argues about the equitable division of the plenty it should enjoy, but the soil itself, the only foundation of that plenty, wastes out beneath its feet, wastes and runs through the broken trees and the torn sod.'[114] In the *Press* in 1940 Lance McCaskill, who would later publish *Hold This Land: A history of soil conservation in New Zealand*, discussed erosion and the role of native flora in its prevention.[115] Of particular importance were his regular articles for children in the 'Press Junior', which sometimes prompted feedback from adults.[116]

By 1940 the emphasis began to shift, from the importance of planting natives in order to hold on to the soil, to the quality of the soil itself.

CHAPTER 5

Vegetables, compost and the Dig for Victory campaign: 1930–50

The earlier emphasis on growing food at home was refreshed during the Great Depression of 1929–33 as people again turned to their garden to provide sustenance. Many garden historians have assumed that vegetable gardening in New Zealand was likewise revived by the pressures of World War Two. While this may have been true in some communities it was not ubiquitous, as will be seen.

The steadily declining birth rate in New Zealand – apparent since the 1880s –provoked concern among '[m]edical moralists, nationalists and imperialists'.[1] Sir Frederic Truby King, a prominent health reformer and founder of the Plunket Society, expressed concern about 'racial degeneration' – the perceived weakening of Pākehā New Zealanders' physical health – and infant mortality rates.[2] His promotion of nutritional foods was echoed in a number of forums and was supported by a growing understanding of the role of vitamins in maintaining health.[3] Eugenic responses, in the same vein as the 'back to nature' ethos that took tramping parties on healthy jaunts into the mountains, included the promotion of 'protective' foods, ideally grown in home gardens. The messages centred on particular crops, specific growing methods and new soil fertility treatments, including compost.

The *New Zealand New Health Journal*, a eugenicist Christchurch publication aimed at female readers, ran a fruit and vegetable garden-

Girl in garden with chickens, New Lynn, 1920s. Poultry were a common source of meat, eggs and manure.

Auckland Libraries Heritage Collections, JTD-11G-05715-2

ing column by David Combridge in 1927. Although the column did not discuss nutrition, its inclusion in a journal about the nutritional values of certain foods made the point well enough: fresh fruit and vegetables were important for health. Soil fertility, garden hygiene and the link between soil health and 'racial virility' were an important part of this discourse.[4]

So, too, was compost. The war did usher in one significant change in home gardening practice: the hot compost heap replaced the cold 'rubbish heap'. It was the result of the efforts of compost enthusiasts, combined with wartime exigencies. Compost-making was posited as a way of rebuilding soils that had been degraded by poor practices, which in turn had led to food crops with poorer nutritional benefits and thence the 'racial degeneration' the eugenicists thought they were witnessing. The philosophies driving the work of compost advocates were not adopted widely, however. It was another example of New Zealand gardeners adopting practices that suited them, and leaving behind those that did not.

Vegetable gardening in the Depression

The 1930s Depression had a marked impact on home food production. In July 1934 the *Press* reported that in Christchurch, 'since August 1933, a depot had been open continuously for the distribution of seeds, plants, manures, and other requirements to relief workers'; 1163 individuals had applied for seed.[5]

A garden in Gladstone Road, Dunedin, complete with chookhouse, vegetable plot and a concrete path to the rotary clothesline. Undated.

Dunedin City Council Archive, TC33 Series, Public Works, S7 3

A second view of this garden shows a small greenhouse attached to the garage.

Dunedin City Council Archive, TC33 Series, Public Works, S7 5

As the Depression worsened, people came to rely on their vegetable garden – referred to as part of a 'substitute welfare state' – for food.[6] Those who didn't grow their own were frowned on.[7] Long-time Christchurch resident Maurice Staunton recollected this sentiment:

> It was the days of the Depression and things were pretty hard, pretty hard indeed. And one chap, he said to Dad, 'Gosh, you've got some great lettuces. Wish I could afford to buy seed.' ... Well, at that time he drew out of his pocket a packet of Temple Bar tobacco and he proceeded to fill his pipe, and Dad said, 'Well, with the cost that you've bought that tobacco for you could have bought all the seed you wanted ... for the garden ... It's a case of rolling up your sleeves and getting stuck in ...'[8]

The spirit of sharing was common, and those who lacked a green thumb were considered fortunate if they had friends who grew plenty. J. Shaw's 1930s letters to Mr Reed of Whāngārei illustrate this generosity. Reed sent boxes of fruit to the Shaw family in Mt Eden, Auckland; his oranges and grapefruit were 'fully appreciated' and his feijoas 'delicious'. Shaw lamented his inability to 'grow some sort of exotic delicacy with which I could repay your generosity; but my garden seems to produce in profusion only pests and weeds'.[9]

With the passage of the Social Security Act in 1938, the government began providing 'sustenance' – a vernacular term for the new unemployment benefit. By 1940, however, government sustenance was inadequate. A productive vegetable garden was one obvious way to combat the hard times. Margaret Jones, who was born in 1920, recalled life at Himatangi during the Depression: 'we had plenty of food, but not much money'. The family made blackberry wine and grew peaches, along with lots of silverbeet, lettuce, pumpkin, carrots and other vegetables, all boosted with plenty of manure from their chickens.[10]

Campaigning for vegetables: 1930s–43

A connection between healthy food and 'racial hygiene' was promoted in the 1930s and 40s by groups such as the Sunlight League and the New Zealand Women's Food Value League. The Sunlight League was a Christchurch-based women's organisation inaugurated in 1931, whose stated aim was 'to promote national efficiency by improving national

A Christchurch gardener with his chickens, 1940.
Photo courtesy of David Musgrave

health'.[11] In 1941 league members formed a gardening group, supplying food to children's health camps and to women whose husbands had been called to war.[12] The group also took a strong interest in soil fertility, including the use of bonedust, superphosphate, cow manure and wood ash, and (by 1945) the making of compost.[13]

The New Zealand Women's Food Value League was formed towards the end of the 1930s in the wake of the Depression, at a time 'when low household incomes aggravated existing health and dental problems suffered by New Zealanders'. Its members were 'urban and middle class, and committee members tended to be highly educated'.[14] The league focused on the apparent decline in home vegetable production, the quality of food preparation in the home, nutrition, the price of vegetables and the impor-

Charles Hale works in his vegetable garden in Lake Road, Auckland, 1932.
Auckland Libraries Heritage Collections, T6365

tance of soil fertility. According to its first *Bulletin*, published in December 1937, its conception was 'due entirely to the vision of Dr Guy Chapman', a dentist with strong views on diet and health.[15] Chapman believed that women – who purchased and prepared the food in most households – had 'a special role in educating other women'.[16] He founded the New Zealand Food Reform Society in 1922, and the New Zealand Humic Compost Club in 1941 (later to become the Soil & Health Association). These organisations were motivated by a radical agenda for a new form of society centred on healthy soil: 'a new social order ... firmly rooted in reality'.[17]

The outbreak of World War Two in 1939 certainly helped to focus attention on home food production, discussion of which had diminished during the City Beautiful hype.[18] Christchurch City Council established a Vegetable Committee on 11 May 1942 'to consider ways and means of securing a regular supply of fresh vegetables to citizens, at reasonable prices'. The committee included a number of city councillors as well as members of the women's branches of the Labour Party and the Citizens' Association, the National Council of Women, the Canterbury Housewives' Union, the New Zealand Women's Food Value League, the Christchurch Central Towns Women's Guild, the Home Economics Association and the West of England Society.

Investigation indicated that there was no real shortage of vegetables, but due to 'a shortage of man-power and petrol, and a glut in prices at the market', producers in many cases preferred to plough produce in rather than take it to market. The prices asked by producers appeared reasonable, and although the retailers seemed to be charging considerably more, 'taking into account the wastage, variation in prices and general overhead expenses to the retailer', their prices were not considered excessive either. Nevertheless, evidence presented by the various women's organisations 'showed that persons on the basic wage could not purchase sufficient vegetables at present retail prices'.[19] The Christchurch Fruit and Produce Brokers' Association maintained that prices were always high at that time of the year, and noted that cabbages and cauliflowers had been 'extremely cheap for some months now'.[20]

While concern about the war's impact on food supply had stimulated the creation of the Vegetable Committee, this was ultimately not the reason for the committee's continued work. If any argument won the day, it was that of the New Zealand Women's Food Value League. Taking their case to the Vegetable Committee in early July 1942, the Christchurch branch of the league pleaded 'the case of the vegetable ... it is a work of national and civic importance to maintain supplies, no matter what the cost or difficulties of money or effort'. In terms of health, they argued, vegetables occupied 'a position of the greatest protective value'. As a result of the drop in vegetable consumption, New Zealanders were suffering from tuberculosis, influenza, eye complaints, catarrh and 'faulty nerve conditions', which included 'neuritis, sterility, inability for mothers to give babies natural feeding, digestive troubles, [and] duodenal ulcers'. They stressed the need for men and women who would 'reproduce readily a revitalised race; men and women fit to take their places in the new world order we are at present fighting to achieve'.[21] Vegetables were the key to this national vitality.

The league provided a number of suggestions for growing and preparing vegetables and asked the committee to consider a 'civic vegetable campaign'. Duly endorsed on 27 July 1942, the campaign's aims reflected long-standing concerns about health rather than immediate concerns related to wartime. It would 'encourage the national use of vegetables as protective foods in prevention of disease', and educate people about the 'mineral requirement of soil for proper production of nutritive vegetables', the 'right kind of vegetables to eat for health' and the 'correct preparation and use of these vegetables.[22]

Three sub-committees ensured the campaign had high visibility for maximum impact. Model vegetable plots were created, and there were radio talks, advertisements, newspaper contributions, a pamphlet, demonstration events, luncheon talks and, of course, a garden competition.[23] Motion picture theatres showed slides depicting the 'relative food values of vegetables'.[24] Other promotional ideas included the production of 10,000 handbills, and the use of sandwich-boys 'to advertise the campaign on Saturday mornings or Friday nights – 12 boys to go in procession round the [Christchurch] Square and then branch off separately to parade different blocks'. It is not clear whether these ideas were acted on.

Dig for Victory: 1943

The outbreak of war in September 1939 had seen a large number of New Zealand's younger gardeners rush to join the armed forces and older ones sign up for the Home Guard. Some 100,000 young New Zealand men were effectively 'removed from the domestic scene'. From 1942 to 1944, New Zealand hosted around 100,000 American troops in camp, and total American visits during the period were as high as 200,000.[25] These troops were mostly stationed around Auckland and Wellington.

Anticipating food shortages, the government made a number of attempts to increase supplies. The Services' Vegetable Production Scheme, in which growers undertook to produce enormous amounts of vegetables, came into force in 1942, and state farms were established to cover any shortfall. The Commercial Gardens Registration Bill was enacted in 1943, under which the government contracted commercial growers to provide produce for the armed forces. Ceiling prices were introduced on a wide range of vegetables.[26]

Worried that this redirection of food supplies might cause shortages for householders, in May 1943 the government launched the national Dig for Victory campaign in Wellington, which encouraged home food production. Although supposedly a national campaign, in reality it was mostly focused on Auckland and Wellington. Couched as a patriotic effort, the scheme involved an impressive array of individuals.[27] Guy Chapman opened a community vegetable garden at a transit camp in

Gavin Paterson (bending) with his daughter Eunice in the vegetable garden, Dannevirke, c. 1943.

Photo courtesy of Lachy Paterson

Gavin Paterson in his vegetable garden, Dannevirke, c. 1943.

Photo courtesy of Lachy Paterson

Western Springs, demonstration plots sprang up in public gardens such as Auckland's Albert Park, and factories, hospitals and other institutions were 'encouraged to plough up their lawns and plant vegetables'.[28] The public was directed to listen to North Island radio stations 'for practical instruction', and stores there reported record sales of seeds and seedlings.[29]

In Christchurch the city council's programme prevailed, championed by the Women's Food Value League. The council and organisations such as the Canterbury Horticultural Society (CHS) provided information on growing produce, and the CHS planned a garden competition to encourage people to greater effort. Superintendent of Parks and Reserves Morris Barnett converted a number of CHS articles on vegetable cultivation into brochures, and suggested that demonstration gardens be developed in suburban parks and the botanic gardens.[30] Radio talks, to be given 'right through the year by garden experts', would provide advice on 'methods of soil cultivation, types of soil, methods of planting, types of vegetables and their suitability for different kinds of soil'.[31]

Dig for Victory committees were established in Nelson and Dunedin in 1944, and at a meeting in Christchurch in September that year Barnett put a motion, later carried unanimously, that a Dig for Victory council be established in the city 'to promote an increase in the number of home gardens, to make ... each home self-supporting in its requirements of fresh vegetables and to provide home gardeners with sound practical advice in the art of planned vegetable production'.[32] The aim of the meeting was to create a new entity to supersede the Civic Vegetable Campaign, which was now over two years old. Notably, none of the women's organisations – so prominent at the beginning of that particular campaign – were represented at this meeting. Barnett explicitly stated that 'he moved the motion on the understanding that the Victory Garden Council would concentrate on the job at hand *and would not go into the nutrition side of the business*, i.e. that the Council would concentrate on the production aspect [emphasis added]'.[33] It was a deliberate capture of a movement that had radical aims by a government propagandist agenda.

The Canterbury Victory Garden Council staged a competition for the best Victory Garden.[34] Seed merchants offered supplies of seed for participants in the competition, and Dig for Victory ideals were listed at public events and delivered to factory workers during lunchtime lectures. The uptake was slow, however. Radio programmes seemed not to be giving the campaign much airtime and, despite widespread advertising, by mid-October 'there had not been any enquiries re the Garden Competition'. The council feared its efforts would be wasted.[35] In the end the competition attracted only 50 entries.

One possible explanation for this public disinterest may be that Christchurch householders already had vegetable and fruit gardens. Interviews conducted with people who were children at the time tend to corroborate this theory. Audrey Potter recalled, 'We weren't affected by that because we had our own ... garden at home.'[36] Judith Todd remembered her mother establishing a productive vegetable garden in 1940; the family called this 'digging for victory' because 'if you didn't dig your garden you didn't get anything to eat!'[37] According to Dorothy Fee, 'everybody used to have their vegetable gardens, and if anybody came to call it was nothing if they happened to have a cabbage under their arm or in the car'.[38] In Cashmere the vegetable garden created by Prue Lovell-

Man Working in a Garden, Christchurch, 1920s or 1930s *by Roland Searle.*
Museum of New Zealand Te Papa Tongarewa, A.019379

Smith's mother was actually *removed* during the war to make way for an air-raid shelter.[39] The well-to-do were perhaps more likely to model digging for victory: Pip Middleton's parents, who owned Lismore Lodge in Fendalton, dug up the front lawn and established a vegetable garden during the war. It was later planted in roses.[40]

This differentiation by wealth is important in understanding our gardening history. In his fascinating 2001 report 'The hidden economy: Vegetable gardening and self-sustenance in Caversham, 1895–1947', James Sagar suggests that vegetable gardening in Dunedin in the 1940s was 'a predominantly working-class affair'. From an examination of aerial photographs from 1947 he concluded that 'households with money in the Caversham area were more interested in establishing the traditional "pleasure grounds" of English high society than with a functional kitchen garden'.[41] The study also revealed a history of 'communal' vegetable gardening in poorer parts of Caversham, where small paths linked unfenced back sections.

Sagar states that the Dig for Victory campaign resulted in the development of a richer vegetable gardening culture in Dunedin, but offers no evidence for his assertion. As in Christchurch, Dig for Victory was only rolled out in Dunedin in late 1944; taking into account the evidence from Christchurch and the paucity of evidence for Dunedin, we can perhaps conclude that this campaign did not have much impact on the quantity of food produced in southern home gardens. Such a conclusion echoes that of Andrea Gaynor, whose study of Australian urban food growing suggests that although many people may have been motivated by the national 'grow your own' campaign, for most it did not represent any substantive change in practice. As in New Zealand, however, it may have legitimated women's existing role in home food production.[42]

Gardening was not everyone's forte, of course. Let us return for a moment to J. Shaw of Auckland, the man who received boxes of fruit from his friend Mr Reed. Shaw was delighted to report to Reed in 1938 that his garden was flourishing – 'much better than I deserve for the labour I have expended … Our roses are a picture & I have recovered my youthful love for the gay nasturtiums', which were 'a thing of splendour, very satisfying to the soul'.[43] In 1942, now at a new address in Mt Eden, Shaw scarcely seemed to be in Dig for Victory mode: 'this summer beat me. My lawn vanished altogether and while that saved cutting it gave us a Saharan aspect … By some sort of miracle I had a good crop of potatoes.[44] In 1946, although he had some success with passionfruit and 'a feijoa I put in in imitation of yours', he confessed he had been trying without success to grow his own oranges: 'Shallow rooting things grow very well – but the deep rooters curl up and die.[45] After Shaw himself passed away in 1948, Reed continued to send oranges to Shaw's wife, who wrote that the friendship had meant a great deal to her husband.[46] J. Shaw was never going to be an avid vegetable gardener, however much he may have desired it.

The compost heap: 'a direct menace'?

During the 1920s and 30s the discourse about feeding soil gradually gathered momentum. J.T. Sinclair, gardening columnist for the *Press* from 1917 34, first mentioned feeding the soil with organic matter in 1920. Supplies of potassic fertiliser had been disrupted since World War One, and Sinclair suggested using ash from burnt garden waste as a replacement.[47] He advocated trenching ('returning to the ground every morsel of material of every sort for the sake of its utility as a direct or indirect fertilising agent'), green manuring with crops 'such as turnips, rape, and mustard', and the application of stable manures and superphosphates. The latter was quick acting 'and should not be applied until spring'.[48] Sinclair was still promoting fertility improvers in 1934: superphosphate, liquid animal manures, horse manure for heavy soil, cow manure for light soil, the digging in of 'plenty of decayed vegetable refuse in the autumn', and 'artificial manures' in the spring.[49]

Jack Humm (the *Press* gardening columnist for 1934–45) regularly mentioned using mixtures of superphosphate, nitrate of potash and sulphate of ammonia to introduce food to the soil. He gave similar advice about organic matter: 'In vegetable gardens where the soil is heavily cropped year after year it is necessary to use animal manure or dig in some green manure crop, such as Cape barley, blue lupin, or decayed leaves to supplement the soil with humus.'[50] In Humm's opinion, humus was 'the life of the soil':[51]

> Without humus no soil can be fully fertile. Mineral elements may be said to give the soil body, but humus gives the soil its life. Without humus bacterial activity cannot develop, and, therefore, the soil is not fully fertile.[52]

The process of using waste materials to form humus created a quandary for some. Sinclair wrote at length on the subject in 1934, stating that fresh manure should be dug in during the autumn or winter and rotted manure – preferably horse manure, which heated up more readily than other forms – applied in spring.[53] Sinclair did not agree with maintaining 'heaps' in the garden: waste material should always be either burned or immediately trenched. Humm was also concerned about garden waste. In 1938 he wrote, 'Do not leave rubbish heaps about in the garden, as they harbour all types of insect pests.'[54] By 1940, however, he had begun to acknowledge the place of the 'refuse heap' in the garden, but warned readers of the dangers of incorporating unhealthy material: 'Keep the garden sanitary by promptly burning all diseased foliage and rubbish. Rubbish harbours harmful insects and fungus pests ... Keep the refuse heap sanitary by dusting hortnap freely after each lot of rubbish is added. Hortnap not only destroys disease but prevents it from beginning ...'[55]

Some readers feared composting would encourage unwanted wildlife. 'Swatter', writing to the *Christchurch Star-Sun* in 1943, asked whether compost heaps might attract flies.[56] A few days later he wrote again: 'Epidemic disease is possible, the fly is a real danger; and known sources of its production should not be allowed to exist. Compost heaps are a direct menace and should be treated accordingly.' Compost piles were 'not essential in the past and a good garden can be had without their aid, if elbow grease is not spared and the hoe is used freely'.[57] W.T. Wainman, a member of the Canterbury branch of the New Zealand Humic Compost Club, thought that old-style, slow-rotting rubbish heaps were a problem: 'Ninety per cent of backyards have one of these.'[58]

Swatter's outcry is indicative of the growing conversations about creating humus, which by 1943 had become a matter of great discussion among garden experts (and the farming community) both in New Zealand and throughout the British Empire.[59] The Compost Club, one of Guy Chapman's organisations, led the charge, promoting the 'hot composting' system developed in India by Sir Albert Howard, a principle figure in the early organic movement within the empire. Advocates contrasted their scientific process of making compost with the old-style heaps Humm was writing about: 'Compost in the old days was obtained from the rubbish heap, but the rubbish took a long time to rot. Composting is a much better and quicker method.'[60] (Humm may have changed his approach later, as in 1950 he gave an address to the New Zealand Humic Compost Club.[61])

T.D. Lennie, president of the Canterbury branch of the Compost Club, took over the *Press* column from Humm as the war in Europe ground to a halt. For some time the club had maintained its own column in the *Press*, penned by committee members on a roster basis under the pseudonym 'Humus'.[62] 'The great improvements noted by scientific agriculturists of incorporating compost in the soil have been so startling that some persons have become possessed with the idea that compost has some mysterious or magical quality,' wrote one. Perhaps this was true, said the writer – before adding, rather radically, 'Most of our social ills can be traced back to the soil.' Perhaps compost could solve humanity's problems as well as producing better veggies.[63]

Mainstream gardening publications also advocated the Howard compost system. The 1943 Reed pamphlet *Home Garden Fertilisers* stated that it 'may well prove the best method ... it has been applied with great success in many different parts of the world'.[64] Howard's system appeared in government publications as well. Although his name was not used and some of the details vary, the characteristic layering of ingredients and the resultant generation of heat is clearly derivative.[65]

The scientific compost heap underwent rapid revision. By 1944 expediency was gaining ground over quality: 'Compost heaps may be either simple or complex,' a government guide offered.[66] In mid-1945 the Canterbury Compost Club suggested taking a relaxed approach: 'great results can be obtained by following less orthodox lines. The great thing is to use, instead of wasting, the organic material that is available.'[67]

There is no doubt that during World War Two a significant shift had taken place in the minds of New Zealand gardeners. The *Press* gardening columnist rejoiced in autumn's falling leaves and added, 'All seasons should be compost time.' Alas, he wrote, one found 'too many fires, in both the country and the town, destroying much that could be made use of incorporated into the soil. When we burn vegetable growth the greater proportion ... goes up in smoke and only the remaining 3 per cent returns to the soil.'[68]

At Waihi Bush near Geraldine, Ethel Musgrave's post-war garden was focused on food. William Turton had died in 1926; his wife Marion remained on the property until her death in 1946, after which the farm passed to her great-niece Ethel, an avid gardener whose diaries illustrate her delight in producing fruit and vegetables. In January 1947 Ethel wrote of picking raspberries and giving away broad beans, lettuces and carrots.[69] In February she planted out cabbage and celery seedlings and weeded her carrots, and visitors in March were treated to bowls of raspberries and currants.[70] When opossums and rabbits invaded her garden, she improved the fences and got on with earthing up the potatoes.[71] In June Ethel made a border; in July she bought new blackcurrant plants, seed potatoes and peas; and in August she 'pruned apples till dark'.[72] In spring she planted a full vegetable plot again. Her garden was not without flowers – Ethel grew snapdragons, mignonettes, asters, nasturtiums and eschscholzias, but seldom referred to them.[73] In contrast to her great-uncle William, she didn't once mention attending to the lawns.

Just when the practice of composting was introduced first at Waihi Bush is not clear. In 1904 William Turton wrote of burning rubbish and using stack bottoms and manure in his garden; in 1947 Ethel Musgrave bought sulphate of ammonia and, significantly, 'did Compost heap'.[74]

Gardener Margaret Jones became interested in compost in 1941 when she met Guy Chapman. At the time, she recalled, compost-making was thought to be an activity for cranks; she recalled a meeting in Auckland Town Hall at which Chapman built a compost heap on stage. Margaret's parents were communists, and as well as absorbing their views she took on the ideas of the composting movement, which were considered almost as radical. When she and her husband moved to a section in Wanganui they 'tried to be self-sufficient'. Fruit trees were already established on the property, and they grew tomatoes and plenty of other vegetables. Although they had a flower garden and a tennis lawn, Margaret stressed that her gardens were always 'productive'. 'You have to get something from them,' she stated emphatically.[75] Margaret joined the Wanganui Compost Club and recalled that branches of the club sprang up all over the country.

Compost turns mainstream

Compost clubs seemed the perfect home for critics of the status quo. In Dunedin, one member, who lived close to Herbert Tyrell (son-in-law of Adam and Mary Fraser: see Chapter Three), was an advocate for defecating in the garden. 'He and his family just crouched in the garden and did it,' Herbert's daughter Lois recalled of the man they used to call 'Mr Compost'.[76] But despite a perceived element of 'radicalism' in the compost movement, *Compost Magazine* claimed in 1947 that, thanks largely to the society's efforts, New Zealanders were now 'compost-minded'.[77] In a meeting of the New Zealand Humic Compost Club in October 1949, Guy Chapman suggested that the club had achieved its aims and should now be wound up.[78]

By 1950 the compost heap could be spoken of as 'a common exhibit in many gardens', even though one journalist feared they were frequently 'not turned over quickly enough as the modern practice advises'.[79] Dorothea Turner referred to composters as artists, saying 'the new art grew intensely'.[80] Even the Department of Agriculture sang compost's praises. In a department bulletin for home gardeners, A.G. Kennelly, who had been part of the Dig for Victory campaign, wrote: 'In recent years the making of compost in home gardens has become increasingly popular and important.' He assured gardeners that 'the properly constructed compost heap offers a cheap, odourless, and hygenic method by which waste material can be converted to vitally needed humus'.[81]

In part, the compost heap helped resolve the problem of reduced access to animal manure that had resulted from the decline of horse-drawn transport. Mary Fraser of Mosgiel was no doubt typical of her generation: her wonderful 1940s garden was fed with horse manure collected from the street. It was a practice carried over from an era that was disappearing.[82] But the motivating force behind the composting movement largely transcended such practicalities. In 1952 Dove-Myer Robinson, New Zealand politician and environmentalist, president of the Compost Club and later mayor of Auckland, described the national progress of the movement:

> Composting is now an acknowledged and 'respectable' activity, recognised by government departments and the public alike, as of very great benefit to the community. To-day marked a milestone in our progress. In our morning paper, I saw an advertisement inserted by the Department of Health, advising the public that the best way to prevent flies breeding in rubbish, is to compost it in properly made compost heaps.
>
> This is a far cry from those days only a few years ago when the Department of Health publicly condemned compost heaps as being the cause of much fly-breeding. This example of co-operation by a government department is a heartening sign which all of us welcome.[83]

A battle of ideas had been won.

'We must achieve a right relationship to the land'

Yeo Tresillian Shand is one example of the 'radicals' who were attracted to the composting movement. An early member of the Royal Forest and Bird Protection Society, in 1941 in his role as a lay Anglican reader he addressed a Christchurch Deanery conference with a paper entitled 'The Crime against the Land', in which he criticised what he regarded as the pillaging of New Zealand soils for the benefit of British capital. He argued that a dysfunctional British imperial economy had led to the

The exhibition tent of the New Zealand Humic Compost Society, Canterbury branch, Christchurch, 1952.

Photo courtesy of the Soil & Health Association of New Zealand

cheapening of agriculture, a decline in human health, environmental catastrophe and a spiritual fall from grace. He quoted a 1941 report of the Malvern Conference called by the Archbishop of York:

> The existing industrial order, with the acquisitive temper characteristic of our society, tends to recklessness and sacrilege in the treatment of natural resources ... The delusion is that cheapness leads to plenty. But of what use is plenty of rubbish? In the strain for this ghastly cheapness, a man's relationship to the soil becomes almost purely predatory instead of by God's laws a process of symbiosis by which everything that has had life has life again.[84]

Shand identified deforestation and erosion as key issues, a point that derived authority from Jacks and Whyte's 1939 publication *The Rape of the Earth*, and was supported by organisations such as the Forest & Bird Protection Society and the Canterbury Progress League.[85] (*Forest & Bird* had publicised the Compost Club's aims, and strong links were retained between the organisations.[86]) Shand was blunt: 'New Zealand, since its invasion by the Christians a century ago, has been largely transposed from a beautiful garden to a stamping-ground for the hard-faced business man and exploiter. Millions of acres stripped of all it has to give …'[87] Whereas Māori lived in harmony with the earth, holding land in communal possession in a 'sacred trust', Christians (by which he presumably meant Pākehā) had 'out-gothed the Goths' in rude vandalism.[88] 'Magnificent forest' had been 'ruthlessly destroyed'; around Gisborne, for example, much of the hillsides, along with English grasses and sheep, had collapsed into the sea, 'And into those same wide open spaces of the Pacific a lot more of our adopted country is being pushed in our blindness – the soil we invoke the God of Battles to help us defend.'[89]

Shand offered a eugenicist criticism: 'Britain gets cheap food for her industrial population, [and] ruins her farmers and the soil they cultivate by its importation … nearly half of Britain's population have only the scantiest access to meat of any kind. Eleven million were actually on a food basis below the minimum requirements of the Board of Health.' The effects of this he claimed were visible at Dunkirk, where the defeated British soldiers were a 'skinny, under-nourished, under-developed lot, with every third man wearing glasses, and the Nazis [were] healthy and bronzed, well-fed and developed to the last ounce of physical fitness'.[90]

The author of an article in the third issue of *Compost Magazine* provided a statement of intent: 'Since the war, and curiously enough only since the war, we have heard a great deal about a "New Order", which is to come after the war. What that New Order is like will depend upon what we – the people – insist on having … In anything like a common-sense civilisation the production of 100% health-giving foodstuffs would be put before everything else.' Ensuring this 'would mean a revolution on the land, for it seems that healthy food cannot be grown by a commercialized agriculture with money as its only standard of value'. The writer suggested that in order to produce better food 'we shall have to go through with whatever changes are necessary to make the soil healthy. There is no other way.'

> It may mean reducing and eventually prohibiting the use of artificial chemical fertilisers. It may mean that the community will have to resume greater power over the land. It may mean dispos-

sessing – as painlessly as possible – people whose sole interest in the land is to draw money from it ... We are all in this together, for no class escapes the results of having half-dead food grown on a half-dead soil.[91]

Radical changes to land ownership or laws governing usage were clearly on the agenda. In a review of H.J. Massingham's *This Plot of Earth*, the reviewer described the author as a 'gardener–philosopher'. Massingham saw that 'big business, science (so-called) and bureaucracy are in alliance against small ownership and individuality'. But more important was his spiritual resilience: 'Though the hosts of God may be in the background [Massingham] still believes that they *are* encamped around the dwellings of the just.' It was still possible for 'man [to] so order his little plot of earth that he can at any time lift up his eyes from it and see the Gates of Eden'.[92]

There is a close relationship between the stream of ideas enshrined in this material and contemporaneous discussions occurring within the Church itself. The National Council of Churches in New Zealand produced a series of pamphlets during World War Two dealing with precisely these issues. Brian Low's pamphlet 'Land and People in Christian Order' condemned the commercialisation of agriculture, which had endangered 'the vulnerable soil'. Forest clearance that resulted in disasters such as the 1938 Esk Valley (Hawke's Bay) floods – where homes were wrecked and farmland inundated with silt from the eroding hillsides – contributed to the familiar refrain about 'the disorders that have arisen both from ignorance and from the absence of Christian purpose, and why we must achieve a right relationship to the land'.[93] Low called for land nationalisation and a reorganisation of cities. In a sentiment echoing Ebenezer Howard's Garden City idea, he wrote that cities could be 'full of light and air and space if we care enough to make them so ... Christians must insist on the importance of human beings and their healthiest development as being above all the claims of money measures ... All the things which may be wanted in any back-to-the-land movement must also be available to those who are *not* farmers.'[94]

Rural Māori gardens: 1930s–50s

This sort of thinking had its effect on Māori as well, 91 per cent of whom still lived rurally in 1936. Land schemes from the 1930s were altering the nature of rural Māori living, and neat cottages were beginning to replace the broken-down shacks that many inhabited. But the disparities between Māori and Pākehā living standards are evident in a study of Ōtaki, a community north of Wellington, published by E. and P. Beaglehole in 1946. They noted that whereas just 2 per cent of Pākehā-owned houses in the community could be classed as dilapidated, 65 per cent of Māori homes were 'beyond repair'. One they described as 'a two-room dwelling, the wood so rotten that a mere touch with a hammer makes a large hole in the weatherboards'. Pere, the man of the house, 'does not bother to garden because he gets his vegetables from the Chinese gardens. His section in any case is too small to bother about flowers.'[95] A second home had both a flower and a vegetable garden; a third had a 'yard full of rubbish, broken pots and pans; broken fences, gates; flower and vegetable garden full of weeds'. Yet another had '[n]o garden outside, ground muddy and dirty, fences broken'.[96] Some members of this community did keep gardens: the next had a big flower and vegetable garden, and another a '[p]leasant flower garden, big vegetable garden'.[97] But for many Māori in this community, the available vegetables were pūhā, watercress and turnip-tops, gleaned from the paddocks of the Chinese market gardeners (who, in many parts of the country were important contributors to the Māori economy[98]). In summary, '[t]he Māori family that has a well-stocked and well-grown garden of vegetables seems to be uncommon'.[99]

The 1952 film *Tuberculosis and the Maori People of the Wairoa District* emphasised the health benefits of 'modern' homes and the value of maintaining a productive vegetable garden.[100] In the film, Māori families in a community in Te Rēinga are living in cramped, run-down dwellings and stricken by tuberculosis. One family, however, has constructed a new home, set amid a sea of dahlias and with a large vegetable patch. At a time when Māori were increasingly moving into urban areas, the film's emphasis reads as an example of taking Garden City thinking to the country.

Nevertheless, many rural Māori still maintained traditional routines in their gardening activities. Honiti Apiti (whose cousin was the daughter-in-law of Hare Puke, mentioned in Chapter One) lived and gardened at Kāwhia during this time and passed on his gardening knowledge to his nephew, Wiremu Puke. Honiti used traditional Māori garden tools and referred to them by their te reo names: he used the ketu, a small paddle-shaped tool for digging kūmara, and kept the timo and a small kō for digging taro, eschewing steel implements because they bruised the vegetables. He tilled the soil from June to August to dig in the frost and 'kill the bugs'. When the kōwhai flowered in September he would bring his kūmara tubers out of storage and start growing the tipu, which would then be planted facing east towards the rising sun.[101] Stories like this provide a useful illustration of the way that gardening traditions may be integral to a sense of identity.

Town waste and triumphant compost

Ebenezer Howard had factored the disposal of town waste into his original Garden City plans, in which he instructed that waste materials should be utilised to manure productive land.[102] Sir Albert Howard, one of a generation of scientists inspired in part by the Chinese horticultural tradition, championed this idea in his 1940 treatise *An Agricultural Testament*:

> The garden city and water-borne sewage are a contradiction in terms. Water-borne sewage has developed because of overcrowding and the absence of cultivated land. Remove over-crowding and the case for this wasteful system disappears. In the garden city there is no need to get rid of wastes by the expensive methods of the town. The soil will do it far more efficiently and at far less cost. At the same time, the fertility of the garden city areas will be raised and large crops of fresh vegetables and fruit – one of the factors underlying health – will be automatically provided.[103]

The much-loathed Chinese practice of using liquid human manure on horticultural soils was now to be embraced. Sir Albert envisaged that the transformation of urban living through garden city planning would start in the colonies where there was ample space. In his correspondence with Dove-Myer Robinson he was optimistic that New Zealand would lead the way.

By the mid-1940s the composting movement had gained traction in Wellington, despite the protests of those who felt that compost contradicted the need to create sanitary city spaces. Robinson wrote to Albert Howard in September 1947 to tell him of the progress since the last election, and in particular about the new Minister of Health, Mabel Howard, who, he said, 'was behind us 100%': 'she has always hated the idea of using artificial fertilizers and she is a great believer in organics ... other members of Cabinet are becoming convinced that it is not only a question of utilising organic wastes, but a question affecting the health and economic welfare of the whole community ...'[104] Ben Roberts, Minister of Agriculture during the war, was another advocate of composting. His widow wrote of his firm conviction that 'as health begins in the soil, we must accept our responsibilities as individuals and help to maintain its fertility by returning to it all organic wastes and reduce our use of chemical fertilizers'.[105]

As the 1950s dawned, the push for composting waste had begun to morph into an 'organics' movement, and remained such into the early

Children work with their mother in an extensive Glen Eden (Auckland) vegetable garden, 1950.

Auckland Libraries Heritage Collections, JTD-12A-05618-3

twenty-first century. New Zealand gardeners had not discarded their spraying apparatus, but compost-making, more or less along the lines developed by Sir Albert Howard, had become a fixed element in New Zealand gardens.[142]

CHAPTER 6

The rise of toxins in gardens:
1920–80

It may seem presumptuous to bracket off 1920–80 as a time of high chemical use in New Zealand's home gardens. Some may protest that gardeners and farmers in the nineteenth century employed worse substances than those introduced in the middle of the twentieth century, but there is no doubt that the use of chemical products to kill plants and insects and promote soil fertility escalated after World War One, and especially after World War Two.

Mary Hobhouse wrote from Nelson in 1861 of an 'exuberance of animal life in the shape of slugs, caterpillars, green fly & blight of every kind' in her garden, but did not mention any chemical controls for these.[1] And there is little for environmentalists to complain about in the 1865 *Hay's Catalogue* direction to gardeners to 'starve out insects' in the field. The recipe for the 'preventive wash' for American blight (or woolly aphid) in the same book leaves a little more to be desired, however:

ılb sulphur
ılb soft soap
2lbs fresh slaked lime
½lb tobacco
½lb red lead
2 quarts of skimmed milk

Once dissolved in two gallons of water, the mixture was only to be painted onto affected parts of the tree.[2] Directions to use 'artificial manure' might raise eyebrows until we see that bone dust was preferred. Lime could be sprinkled around onions 'occasionally' to keep slugs down: the best time was 'a quiet still night about nine o'clock'.[3]

Mid-nineteenth-century gardening directions for chemical interventions were generally on a small scale, and most products were relatively benign and usually targeted. In Britain and Europe the application of chemistry to agriculture was only just beginning, notably with Justus von Liebig's 1840 work *Organic Chemistry in its Application to Agriculture and Physiology*; Rothamsted Experimental Station was established in England in 1843 to research and disseminate information about chemical fertilisers.[4] And although farmers were using more artificial pesticides than before by the end of the nineteenth century, the options available were limited. In New Zealand, the Orchard and Garden Pests Act 1896 signalled a change in approach, requiring of orchardists a compulsory annual dressing of fruit trees to prevent the spread of 'apple-scab' and codling moth, and 'at all times [to] do whatever is necessary in order to eradicate phylloxera [an aphid] ... and prevent the spread of that disease therein'.[5] By the early 1920s, approaches to disease and pests were becoming virulent.

The war against insects

According to the *Press* in 1920, although World War One was over, 'the war against the insect world was only just starting, and it would prove to be a more deadly fight than any men had had before'.[6] How accurate this claim was. As we saw in the previous chapter, the perception of flies as vectors of disease was a barrier to the introduction of compost heaps in gardens. 'There was battle to be waged against the deadly blow fly, house fly and mosquito, the Huns of the insect world,' the *Press* went on.[7]

Newspapers began to promote the use of sprays in home gardens. J.T. Sinclair, author of the garden column in the *Press*, in 1920 recommended a mixture of air-slaked lime and tobacco dust for caterpillars, and dilute Black Leaf 40, a nicotine preparation, for woolly aphid – 'a serious enemy'.[8] At least the latter had a limited environmental impact. A benign soot mixture was suggested to keep flies off vegetables and flowers; in extreme infestations, kerosene was useful.[9] Soot would also keep slugs off celery.[10] The *New Zealand Herald* suggested soot water and lime as part of the remedy for clubroot in cabbages in 1919, and Bordeaux mixture (a combination of copper sulphate and slaked lime) or lime–sulphur mixtures for fungal problems in peaches.[11]

But attitudes towards pests were changing. An article in the *Press* in 1920 heralded the shift:

> The greatest drawback to fruitgrowing … in the suburban garden … is the ceaseless toll which insect pests impose upon the trees. Woolly aphis and codlin moth on the apple, black aphis, fruit fly, and tip moth on the peaches, and scales – brown, olive, red, Indian wax, and others – on the citrus … are an ever-present reminder that only by eternal vigilance can the fruit be retained in a suitable condition for human use.

According to the *Press*, 'The spray pump and the fumigating tent [for cyanide applications] are offered as means of eradication', but, the writer asked, 'how many men have the time and patience to use them effectively?'[12] A new technique – of drilling pesticides straight into the tree so that poisons were delivered directly to pests via the sap – could be more efficient.

Pesticides got nastier. Calcium arsenate is a case in point: by 1999 it would be listed as a known carcinogen responsible for a host of serious disorders, and the handling of calcium arsenate would require full protective gear with breathing apparatus.[13] But in March 1922 one newspaper described how to prepare the spray, by dissolving a quantity of sodium arsenate in water, slaking lime in water and combining the two. The resultant paste could be strained through a cloth to eliminate lumps that might clog the spray nozzle. 'Calcium arsenate, like lead arsenate, will not burn the foliage and is considerably cheaper than this poison, which has been favoured so much of late years as a substitute for Paris green.' As a remedy for 'potato bugs' it had apparently been used 'with unqualified success'.[14]

This little case study is a helpful entrée into the world of toxic chemical applications in our gardens. 'Paris green' (copper acetoarsenite, now regarded as 'extremely toxic'[15]), in use internationally from the 1870s through to about 1900 as an insecticide, was succeeded by lead arsenate, which at least had the advantage of not burning foliage.[16] These preparations were used for potato beetles and, particularly in the case of lead arsenate, for codlin moth. World War One saw a rise in lead prices, however, and resulted in an increase in the application of calcium arsenate after this point.

Each of these was highly toxic in its own right. Although this was understood – Francis Peryea of the Tree Fruit Research and Extension Center, Washington State University, mentions that arsenic residues were found on produce sprayed with lead arsenate in 1919 and alternatives sought thereafter – lead arsenate in particular remained in use in New Zealand until the 1960s.[17]

Globally, however, after World War Two lead arsenate was largely replaced by the inorganic insecticide Dichlorodiphenyltrichloroethane, commonly known as DDT, which was created in 1939. DDT was produced in New Zealand at the Fruitgrowers' Chemical Company in Māpua from the 1940s, along with other chemicals like dieldrin, paraquat and 2,4-Dichlorophenoxyacetic acid (2,4-D). (By the time it was closed down in 1988, Māpua was the most contaminated site in New Zealand.) A glance through gardening guides, particularly those of the 1950s, suggests that DDT was in wide use by home gardeners for the dispatching of various grubs. *Te Ao Hou*, a publication aimed at Māori, advised in 1955 that apple trees should be sprayed for codlin moth every two weeks using DDT or arsenate of lead.[18]

Numerous studies have pointed to the toxicity of DDT to humans, birds and aquatic life. It persists in soils and bioaccumulates up the food chain. The New Zealand Healthy Life Society in 1970 alerted members to the dangers of DDT, but just how widely this message was received is unknown.[19] DDT was eventually banned in the United States in 1972 and from use on New Zealand farmland in 1970, but it remained commercially available for home gardeners here until 1989.

A less pernicious chemical, and one that was allowed under the organic standards of the 1980s onwards, was Derris Dust, a product derived from a natural plant source that contains the crystalline substance Rotenone and which is used as a pesticide. Of all the chemicals commonly used by New Zealand's home gardeners of the mid-twentieth century, Derris Dust may be the best known. It was used to treat sheep lice early in the century, and it appears that use expanded to treat other pests internationally from the early 1930s.[20] Fortuitously perhaps for New Zealand growers, that development coincided more or less with the arrival in 1930 of the cabbage white butterfly. Commercial growers of brassicas had become accustomed to the insecticide by the early 1940s, at least in Australia, where its primary use was against green aphids.[21] Whether Derris Dust was in wide use by home gardeners in New Zealand this early is unclear; it wasn't mentioned in the *Press* gardening columns of 1940, but was recommended as a control against 'moth grubs' on cabbages and cauliflowers in 1950.[22]

The jury is still out regarding the safety of Derris Dust; following the discovery of a possible link with Parkinson's Disease, the Soil & Health Association campaigned to have it removed from organic production in 2001.[23] But the use of Derris Dust was firmly entrenched among home gardeners. One gardening writer noted in 2007, 'Have your Derris Dust or Pyrethrum spray ready to wage war on the white butterflies. They love all the Brasicca family, Cabbages, Caulis etc plus other tasty plants.'[24]

The uptake of new chemical applications for the home garden seems to have been relatively slow until around the 1950s. Along with lead arsenate for insects and various sulphur concoctions for fungi, Jeyes Fluid was recommended in the 1930s (for grass grubs), as was red lead (to keep sparrows off lawn seed), horticultural naphthalene (hortnap) in the 1940s to 'maintain a clean garden', and a substance called 'Garden Flit' which promised to kill insects and fungus pests. Supposedly not poisonous to humans and 'animals', it was 'sure death to garden pests'.[25]

Albert Paynter kept his prize-winning Fendalton garden up to scratch during the 1940s with a lawn fertiliser and sulphur-based sprays for the roses; another Fendalton family used Black Leaf 40.[26] But ordinary gardeners, especially those with limited financial means, did not always resort to these products. Herbert Tyrell, who lived in Brockville, Dunedin, from the late 1930s, was one such gardener. He kept chickens and had a bountiful garden with 'virtually ... every vegetable you could grow in the South Island'. Herbert's daughter recalls that her father, and indeed most people they knew, never used sprays. Slugs and other insects were largely dealt with by the chickens, which were let out to tidy up the garden.[27] Herbert also had a huge compost heap. Instead of artificial fertilisers, chicken manure was soaked in a drum of water then watered into the garden. He also made use of human waste from the family's bucket toilet.[28] This was probably a fairly common practice; one correspondent to the Christchurch *Press* asked about using night soil in the vegetable garden in 1930 and was advised that it could be used – but 'sparingly'.[29]

Some attitudes to chemicals for the home garden as expressed in gardening guides of the 1950s are alarming by today's standards. The Yates guide for 1957 suggested that poisons for chewing insects should be applied to 'cover the whole of the plant'; repeat sprayings would be necessary 'to ensure that the new untreated growth receives its coverings of spray'. The sprays, the guide assured its readers, 'may not necessary [sic] be toxic to human beings', which was no doubt a relief. DDT and Derris Dust were described as 'harmless to us but a poison to many leaf-eating insect pests'. 'On the other hand,' the guide continued, 'Arsenate of Lead, which was frequently used before the advent of DDT, is a deadly poison.' It is fascinating and slightly disturbing that, on the very same page, arsenate of lead is recommended along with DDT and Derris Dust as a control for caterpillars.[30] The guide also suggested Paris green as a remedy for crickets, grasshoppers, snails, slugs and slaters.[31] Formaldehyde, now known to be a respiratory irritant, was a useful soil steriliser; the soil should be 'thoroughly soaked' in it.[32]

Organics and chemicals: complementary or not?

Many people thought chemical sprays and fertilisers complemented 'organic' forms of gardening. Jack Humm, who was passionate about humus and the complex webs of soil life, stated in 1940 that for strawberries, 'hortnap' should be dusted 'freely in the bottom of the trench [to] destroy fungus and insect pests' – an intriguing contradiction.[33] T.D. Lennie, president of the New Zealand Humic Compost Club from 1954, advocated for the use of superphosphate in the 1940s.[34] In 1950 he described fungal blights as 'deliberate destroyers' comparable to 'such disgusting animals as rats', and recommended early spraying as 'more a preventative than a removal'.[35] The 1955 *Star Garden Book* presented what was perhaps a majority view on the matter: a less vigorous advocate for poisons than the *Yates Garden Guide*, this manual promoted the importance of soil fertility:

> The basis of successful gardening should be the maintenance of
> an ample supply of humus by the regular use of compost, [but] the
> addition of necessary plant foods in more concentrated form can
> aid in building up fertility and in giving a greater return of produce
> from a given area.[36]

To control insects it recommended Paris green, arsenate of lead and the 'recently developed insecticides' HETP (hexaethyl tetraphosphate) and Lindane (*gamma*-hexachlorocyclohexane) – both to be handled with care.[37]

(The *Yates Garden Guide* instructions for making compost are worth noting for their contradictory advice. Compost was 'indispensable to every home gardener'; it was 'the law of Nature that everything of organic growth, having lived its life, dies, decays, and returns to Mother Earth to enrich the soil for the benefit of the living plants to follow'. On 'no account should any wood-preserving substance' be used on the boards of a compost bin, as this would contaminate the compost. The guide then advocated the addition of superphosphate and sulphate of ammonia to the heap – something many composters would find abhorrent.[38])

Gardening with science: 1950s

The use of this chemical arsenal in the war on insects, weeds and disease relied – in one view – upon turning gardens into 'war zones', where every unwanted plant or insect was treated as an 'invader' requiring elimination. In hindsight it may seem incredible that by the 1950s many gardeners did not believe a decent garden *could* be maintained without these lethal weapons. But this view ignores the many pressures on householders, the most significant of which may have been the need to conform in the post-World War Two suburban landscape. 'New', 'scientific' and 'safe' gardening aids were warmly received.

The *Otago Daily Times* gardening column, introduced in 1952, was aimed both at those who 'cultivate their plots with enthusiasm' and those who gardened 'merely from a sense of duty'. With a focus on 'modern gardening', the writer promised to abolish 'time-wasting, obsolete methods' by showing readers 'how to use the latest discoveries of the plant scientists'.[39] There was advice on the best chemicals to use for particular problems: for eelworm in the narcissus, the answer was to fumigate with formalin.[40] To improve soil for broad beans, well-rotted compost was recommended – with the addition of sulphate of ammonia and superphosphate.[41] Lawns would also respond well to sulphate of ammonia and superphosphate mixtures, and the addition of 'a little red lead to the seed', and spraying a lawn 'lightly with kerosene after sowing' would discourage birds from disturbing newly seeded areas.[42] What a contrast to the 1920s advice in the *Press*: 'Black cotton stretched across from low sticks as supports will keep birds at bay.[43] In a letter to the editor in 1950, 'Cocky' suggested putting out water for birds laced with 'a good dose of poison': 'The damage that is being done in the vegetable and fruit gardens is simply terrific,' Cocky complained.[44]

One garden diary written in the mid-1950s by a Christchurch child provides some insight into how and why chemicals were used in the home garden. Rayna Wootton's family home had a large lawn out the front surrounded by shrubs and flowers, and a massive vegetable and fruit garden out the back where they grew a range of vegetables along with apples, peaches, plums, grapes, currants, berries and rhubarb. Rayna and her father maintained soil fertility with liquid chicken manure, blood and bone, 'Iva Plant Food' for potatoes and compost.[45] In a patch of her own in 1953 Rayna grew peas, beetroot, onions, carrots, radishes, lettuces, potatoes and cabbages. Her father initially showed her how to dangle strips of paper from string to protect her pea seeds from birds, but these 'did not do much good'.[46] The following year she recorded that he put 'about 2 drops of kerosene in a tin and swirled the peas around in it. This gives them a coat and protects them from the birds,' she wrote confidently.[47] They sprayed carrots with Black Leaf 40, treated fungus growth on the soil with Jeyes Fluid, and dressed the cabbages and lettuces with sulphate of ammonia.[48]

One notable absence in the range of chemical aids adopted in this garden is herbicide. George Archibald Shillito (Archie) maintained an equally abundant garden in Timaru from the 1950s. The only ingredients he added to his garden were compost, blood and bone, lime – which he sprinkled periodically on the ground – and Derris Dust, which he used on the cabbages. His daughters were enlisted to remove slugs by hand. Archie favoured diligence over herbicides. When an acquaintance newly arrived from England asked what to do about weeds in his own garden, Archie told him, laconically, 'Pull them out.'

'But they keep coming back.'

'Yes. You just keep pulling them out.'[49]

Garden competitions continued to set the tone for perfection. One held in Dunedin in 1953 attracted 62 entrants. The judges commented on the high standard of vegetables and compost heaps; some lawns, on the other hand, would benefit from more sulphate of ammonia, superphosphate and '[m]odern hormone weedkillers'.[50] Images of the winning

'Judges in garden', from a photograph album of shows, festival displays & garden competitions (1955–57).

Dunedin Horticultural Society records, Hocken Collections – Uare Taoka o Hākena, AG-524-05/001

gardens, as judged by the Dunedin Horticultural Society (DHS), illustrate the high standards expected. One 1956 vegetable plot is utterly weed-free with no sign whatsoever of insect or caterpillar damage.[51] It is almost inconceivable that this particular garden's ecosystem could have been robust enough to achieve such perfection without recourse to chemical aids.

Compost dissidents: 1950s–1970s

Not everyone subscribed to the new science or ideology, however. The Otago branch of the New Zealand Organic Compost Society (OCS), the renamed Humic Compost Club, insisted in 1957 that 'compost-grown seeds gave better results than those grown with artificials'.[52] The society wrote to all political parties of the time to seek a 'definite statement' on 'compost or artificials', and continued to promote the 'necessity of compost grown food to maintain health'.[53] A 1957 *Otago Daily Times*

'Vegetables (1958)', from a photograph album of shows and garden competitions, 1957–59.

Dunedin Horticultural Society records, Hocken Collections – Uare Taoka o Hākena, AG-524-06/002

'Vegetable garden', from a photograph album of shows, festival displays & garden competitions (1955–57).

Dunedin Horticultural Society records, Hocken Collections – Uare Taoka o Hākena, AG-524-05/002 feature on the branch's development of a 'communal garden in Nairn Street' suggested there was 'plenty of room for difference of opinion about some of the more radical views held by members of this society', but agreed that OCS's basic tenet – that all waste should be returned to the soil – 'is one with which there can be small ground for quarrel ... Science is important to members of the Compost Society.'[54] The 'more radical views' were likely a reference to the group's stance against artificial chemicals. A paper entitled the 'Challenge of 1963', presented to the branch in the wake of the Cuban missile crisis, discussed ways to improve the nutritional value of food, 'thus fortifying our health against the dangers imminent from radiation, etc'.[55] If nuclear war were to break out, those who ate food grown in compost-enriched soil were deemed more likely to survive.

The larger theme here – of two opposing horticultural camps becoming increasingly paranoid in their relations with one another – fits neatly with the times. One OCS member expressed their view as 'ceasing to think on the physico–chemical plane & learning to think on the biologico–ecological plane'.[56] The competing sciences, and indeed competing philosophical systems, were certainly not reconciled during the twentieth century.

Attendance at public meetings of the Otago branch of OCS gradually increased from 15 in 1963 to 50 in 1969.[57] For those not ideologically enrolled in the organic movement's aims, however, composting remained a useful aspect of gardening practice. In 1977, while addressing the Otago branch of the Soil Association (the latest incarnation of the Humic Compost Club, which had changed its name again in 1970), the president of DHS assured listeners that 'Horticultural Society members shared the Soil Association's view regarding the benefits to horticulture of compost, mulch and animal manure, and suggested that all groups interested in gardening should co-ordinate their activities in promoting common goals.'[58] The following year the Soil Association held a stall at the DHS winter show, displaying organically grown produce, food preparation ideas, wholemeal bread, a compost bin, instructions for compost-making, copies of *Soil and Health* (the successor to *Compost Magazine*) and other association literature.[59] This show of togetherness would have helped reinforce the concept that composting and organic gardening were connected parts of the broader practice of horticulture.

Proceeding with caution: 1960s

Most home gardeners do not appear to have bought into either philosophical system wholeheartedly. They adopted certain practices, but not necessarily the belief systems promoted by the main proponents of those practices. An articulate insight into the debate can be found in Cicely Wylie's delightful 1964 book, *A Garden at My Door*. Describing the impact of the 'Albert Howard' compost system and her husband's initial distaste at the thought of the flies and rats it might bring, she explained the hearty debate in her neighbourhood about the right way to make a compost heap and, in particular, her struggle with the self-appointed odd-jobs man. 'Old Tom' was of the old school of trench-making:

> Unfortunately he loves digging. The deeper the better. So I stand helplessly by, while he makes a trench big enough to bury his past. But it's my precious compost he is burying ... I'm a faddist about this composting business ... [It] is no illusion that it increases the flavour of our vegetables and intensifies the colour of our flowers.[60]

Cicely wrote that since reading Rachel Carson's 1962 book *Silent Spring* (which documented the adverse environmental impact of indiscriminate pesticide use), 'we are in a quandary about whether to spray, or not to spray'.[61] Sprays might upset the balance of nature, she thought, but although Old Tom advised her that insects like leafrollers would just disappear in time, nevertheless she found she could not tolerate caterpillars and other 'greedy, nibbling things' on her precious plants.[62] 'So I suggest ... spraying the flowers and leaving the vegetables. Though this is really no solution, at least I'll have roses without greenfly.' The spray might drift onto the food crops, but it was a risk she was prepared to take.[63]

Len Paterson maintained a large vegetable garden from early 1962 on a quarter-acre section in Whakatāne. The Patersons' home was a working-class house bought from the local paper mill. The front yard was mainly in lawn with a few shrubs, which his wife Sylvia maintained 'just to keep things looking respectable'. The back section, typically, was almost entirely given to vegetables and fruit trees. There was a large, double compost bin made of concrete blocks into which everything went, from garden waste to kitchen scraps, including chicken bones and, at Christmas, turkey skin. Rats lived in the heap, but the family cats kept these in check. The toilet was initially 'a can' in the garage. Until the house was connected to the sewerage system in June 1962, toilet waste was collected by the municipal night soil collection, although his son Lachy remembers one occasion when Len dug a hole in the garden to empty the can.[64] (Lachy also recalls that when he lived in Palmerston in the early 1980s, 'quite a few people in the town buried their crap on their sections'.)[65] Composting did not necessarily indicate an acceptance of organic principles. Derris Dust was used to fend off cabbage white butterflies, Slugslam to deal with slugs, and Len occasionally mixed a copper spray for the fruit trees or roses.[66]

A similar approach can be seen in the Hastings garden of Iris Keenan, established in the 1930s and maintained by Iris after she was widowed in 1961. Iris relied on deliveries of cow and sheep manure from her farming son-in-law Robert Tankersley, applied blood and bone to her roses, and sprinkled lime around the garden. For pest control she used Derris Dust on the brassicas but little else. Her daughter and Robert kept a rural garden at nearby Maraekakaho in much the same style, although in addition to Derris Dust, Robert sprayed copper sulphate on the roses and kept aphids at bay with soapy water. Rather than making compost, the couple simply collected cow manure and left it to rot before applying it to the garden.[67] Robert's parents Winnifred and William (Tim) Tankersley lived next door. In the 1960s their extensive garden boasted grapes, an old walnut tree, a feijoa and a venerable quince tree. This couple's chemical inputs were also low: weeds other than blackberry were dug rather than sprayed, and Tim was 'cautious about agricultural pesticides' and used DDT sparingly on the farm. Winnie resisted any chemical interventions in the garden, used netting to keep birds off the grapes, and tried to get to the quinces before the possums.[68]

Compare this to the garden at Waihi Bush. We have already seen how Ethel Musgrave used sulphate of ammonia in 1947, and we know she had a visit from the 'DDT man' the same year.[69] Her brother Max took over the farm in 1957. During the 1960s Max kept a massive vegetable garden that included 'a big patch of spuds and carrots and brassicas', 'lots of beetroot' and 'a big asparagus patch'. Although he used 2,4-D and 'DDT Super' on the farm, in the garden itself he carried out most of the weeding and pest control by hand. He used winter oil (a mineral oil to kill scale insects and lichens) and Bordeaux mixture on the apples, and kept a supply of Black Leaf 40. For fertiliser Max mixed superphosphate and blood and bone and put 'a small handful ... under each spud' at planting time.[70]

The Musgraves' garden is reminiscent of the earliest gardens made by European settlers in New Zealand in content, especially in its emphasis on potatoes and asparagus – even to the presence of a cherry house. Some of the chemical applications, such as Bordeaux mixture, were unchanged, but recent innovations such as insecticides, compost and artificial fertilisers had now been introduced. The new arsenal for the garden was considered so ordinary and 'safe' that it did not register as being the same as 'using chemicals'.

Māori gardeners did not always adopt every new chemical recommended to them. Rupert Whihongi (Ngā Puhi) and his wife Winifred (née Collier, Ngāti Porou) lived in Gisborne. Until Rupert's passing in 1978, their 1000m² section was crammed with vegetables. They did not use chemicals or sprays in their garden, and devoted the entire space to food production. Their daughter Mary Jane White (b. 1950) also avoided sprays, as her son Roger later recalled. Peter and Mima Richardson (Tainui/Te Arawa) kept an extensive vegetable garden at Ōhakea in the 1930s and 40s. They used a traditional moon gardening calendar and also eschewed all chemical sprays. According to Darcia Solomon, whose parents Rangi and Miriama Solomon maintained extensive gardens in Kaikōura during the 1950s, chemical applications were considered 'tapu tapu to Māori. It was poisonous.'[71] Her whānau grew flowers among the vegetables to deter insect pests, and maintained soil fertility with sheep manure. They prayed for the success of their crops. Kumeroa Mason, also of Kaikōura, and Tana and Elva Tomoana, of Wairua near Hastings, avoided chemicals too.[72] Even when a kind of wire worm attacked the Tomoanas' sweetcorn and maize they didn't spray – and the problem eventually 'just settled down'. There was no trenching system, and composting was a matter of 'just dumping everything and mixing it up'. Tana and Elva relied primarily on an unstructured crop rotation system to maintain fertility, and in this way grew 'fantastic', 'tasty' and 'prolific' vegetables during the 1950s and 1960s. Tana continued to use the moon calendar his father had written according to ancestral knowledge and was diligent in doing so – possibly because his father helped out on their property.[73]

Conformity, weeds and social diseases: 1950s–1970s

The need to conform was keenly felt by most: people who allowed their lawns to grow too long or their front gardens to be overwhelmed by weeds were deemed to have thumbed their noses at society. Rayna Wootton, who grew up in Fendalton, put it this way: 'It reflected on the householder, really, whether they kept a good garden.'[74] According to Pip Middleton, also of Fendalton, 'You didn't want people looking over your front fence and seeing weeds in the front garden. That was a definite no-no ... [If] you let the street down ... it was worse than death!'[75]

Conformity could be stressful, however. Florence Morris, who lived in the hill suburb of Roslyn, Dunedin, from the 1940s to the 1970s, was quite unable to convince her husband John to help outdoors. According to her son Brian, she sometimes 'begged' her husband to mow the lawn

Ken Morris in front of the broad beans and courgettes in Florence Morris's garden, Dunedin, c. 1952.
Photo courtesy of Lois Morris

but to no avail. 'She tried desperately for a garden', but found the solid clay was too hard to break up. John frequently brought home commercial travellers and groups from the local officers' club, and Florence was 'desperate' to keep the frontage, with its lawn and small edging garden, looking good.[76]

Looking back, it was as if weeds had become a kind of social disease that threatened to overwhelm the veneer of suburbia. Reading through advice columns and garden advertisements of the 1950s and 60s, it is no great stretch of the imagination to see reflected in them the fear that communism could still 'break out' in New Zealand. An unkempt garden was surely indicative of an anti-social community member? According to Brian Gilberthorpe, many Christchurch homeowners were fanatical gardeners: 'Some streets must have been sheer hell to live in if you weren't interested in gardening … You would be ostracised … if you didn't measure up!'[77] Now that modern science was making the work so easy, there was really no excuse for an unkempt garden.

Call to chemicals

The herbicide 2,4,5-Trichlorophenoxyacetic acid (2,4,5-T) was developed in the 1940s and produced in New Zealand from 1960 by Ivon Watkins-Dow at their Paritutū plant in New Plymouth. Production of this chemical created the dioxin 2,3,7,8-Tetrachlorodibenzo-p-dioxin (TCDD), nowadays a known carcinogen. (TCDD would later be combined with 2,4-D to make Agent Orange, the defoliant used in the Vietnam War.) The dioxin caused birth defects, miscarriages and tumours in laboratory animals, leading the US Environmental Protection Agency to suspend the use of 2,4,5-T in 1979.[78] The product continued to be available in New Zealand for another eight years, however, primarily for agricultural purposes.

The Soil Association published a steady stream of articles denouncing the continued use of 2,4,5-T. These spoke of a 'catalogue of horrors' – a reference to deaths linked to use of the chemical: there was 'no reason whatsoever … for there to be any doubt of the damage done by 2,4,5-T and 2,4-D'.[79] Despite the publicity, Ivon Watkins-Dow steadfastly continued manufacturing 2,4,5-T until 1987, at which time it was finally banned in New Zealand. A number of studies undertaken in Paritutū between 1985 and 2001 established that traces of TCDD were present in soils around the factory, although not to any level that would cause health issues.[80] A subsequent 2002 study found TCDD in residential garden soils in the vicinity of the factory at depths greater than anticipated, but a joint statement from the ministries of health and the environment concluded that no further testing was required.[81]

The war on pests increasingly took on the appearance of an arms race as the 1960s wore on. New plants were being introduced from all corners of the earth, and '[d]espite the tremendous advances in chemical weed control, the problem of keeping a garden free from unwanted plants is much more difficult now,' wrote one expert in 1967.[82] Although most people weeded for the sake of 'keeping up appearances', he wrote, there were more important reasons to be vigilant. Weeds encouraged

The front cover of Soil and Health, *Winter 1986.*
Soil & Health Association of New Zealand

pests and diseases in a garden, but even more serious was 'the way in which various diseases, often with the active co-operation of insects, move from weeds to other plants'. Conspiratorial insects! '[I]f we are to grow healthy fruits, vegetables and flowers, systematic weed destruction, particularly in the early stages, becomes imperative ... Even the most harmless looking weed can be the host for a trouble-maker.' The answer, according to this writer, was regular weeding accompanied by the application of chemicals to keep things in check: 2,4-D for broadleaf weeds in lawns, and pre-emergence weed-killers, selective grasskillers and 'complete killers' for weeds on drives and paths.[83] In the same newspaper a prominent advertisement for Jeyes Fluid claimed it would control garden pests and fungus when used as a soil steriliser. It was 'the best thing on earth'.[84] Sterile, clean and with troublemakers all kept at bay, the garden would be perfect!

The most popular selective broadleaf herbicide was 2,4-D, which in the early twenty-first century was present in over a thousand herbicidal products on the world market. It was also contaminated with dioxins, but was not banned as studies into its toxicity to humans were deemed inconclusive. Roundup, a broad-spectrum (i.e. non-selective) herbicide from the global chemical company Monsanto, appeared in 1973, and by the 1980s was in heavy use worldwide. Roundup was promoted as absolutely safe and non-persistent, supposedly becoming inert as soon as it came into contact with soil. Numerous studies undertaken over the years have demonstrated this point, but others have shown a link between use of the chemical and neurological disorders, cancers and other illnesses.[85] Some studies have suggested that Roundup can have adverse effects on ecosystems. A general ambivalence about the chemical prevailed even early in the twenty-first century.

A report by Brian Maunder analysing soil contamination in 12 Auckland residential properties, most located in subdivisions of the 1950s and 1960s, showed that all too often pesticides and herbicides did not break down quickly:

> Many New Zealanders were keen home gardeners ... from the 1940s
> through to the 1970s, and many residential properties had home
> vegetable gardens and home orchards which provided a significant
> source of produce for home consumption. Lead as lead arsenate,

> DDT and formulations of copper were among the range of chemi-
> cals used to control pests in these situations … It is not surprising
> to find domestic sources of horticultural chemical residues on
> long-established residential properties.[86]

Many people found this a matter of great concern. 'It is alarming to read almost every day, the tragic results on the body and mind, of the seemingly sensible use of chemicals in gardens,' the Doubleday Association newsletter announced in 1979.[87] It was but one voice among many.

While there is some evidence to suggest that the organic movement's call to reduce the use of toxins had penetrated mainstream gardening consciousness by this time, in general the chemical companies were winning the debate on chemicals. Although the garden columnist for the *Press* in 1975, M. Lusty, from time to time discussed such things as the value of ladybirds as predators of insect pests, or made reference to a study showing that birds removed about 95 per cent of codlin moth caterpillars in unsprayed orchards, or suggested not spraying apples in order that 'natural predators in the form of birds and other insects would be given more scope', he or she also advocated the use of pesticides and herbicides.[88] Correspondents to the paper frequently mentioned the chemicals they had been using, such as one in 1975 who had used Banvine and 2,4-D on their lawn to kill hydrocotyle (or pennywort, which had become a problem in Christchurch that year). In response, Lusty recommended 2,4,5-T.[89] In 1976 Lusty stated that although 'there is a growing contention in some quarters that really healthy growing plants are able to overcome adversities [such as aphids and caterpillars] with very limited or no recourse to spray control, this view is not shared by this writer'. He recommended Derris Dust for 'those who prefer organic controls', noting however that it needed more applications than other products; he suggested menazon, maldison, carbaryl and diazinon be applied to cabbages and other brassicas; he recommended maldison for gladioli; and, in a reversal of his earlier opinion, advocated spraying fruit trees, adding that '[m]ost commonly available pesticides are compatible with one another and controls can generally be mixed together'.[90]

By the 1970s chemicals of one sort or another were part and parcel of standard gardening practice for many, but by no means for all. The Matoe family, based from 1977 in Philipstown, Christchurch, maintained a hot compost and used Derris Dust but few other unnatural controls.[91] George Shillito's daughter Ali moved with her husband Tom Kennedy into a cottage with an established garden in rural Oxford in 1980, where they maintained compost heaps and gathered sheep manure from under the shearing sheds to dig straight into the garden as a fertiliser. The only pesticide they used was Derris Dust.[92]

Walter Hamilton of Devonport, Auckland, is typical of an older generation of gardeners in the 1980s. His 1987 garden diary records the sowing, planting and harvesting of an impressive range of vegetables, and recounts his interest in roses and cinerarias. Walter typifies the ambivalence within the organic movement about certain sprays and fertilisers. He used plenty of compost, seaweed and blood and bone, yet despite the fact that he was intimately involved with the Soil Association of New Zealand and was an activist in the Devonport compost scheme established in 1979, he was not averse to spraying Roundup on his paths.[93]

What can be seen over this period, therefore, is an increasing call to use chemicals in the garden from the 1920s to the 1980s, and a counter effort against this from what was becoming an 'organic' movement. Many gardeners were selective, opting for affordable sprays and chemicals that appeared less toxic; others advocated tolerance of a little bit of pest damage, and pulling weeds instead of spraying. For most, it seems, a tidy garden – one that was 'properly' kept under control – was a sign of being a good neighbour and citizen. While chemicals offered one pathway towards this goal, another was to entirely recraft the format of the garden itself so that it simply did not require the same level of maintenance. This process of simplifying gardens was well under way from the 1960s, as will be seen in the following chapter.

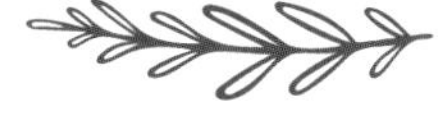

CHAPTER 7

The collapse and renewal of home gardening culture: 1960–2020

For many property-owning New Zealanders in the 1960s the importance of having a garden around the house remained unquestioned. Not all properties had one of course, but those with available space still tended to have vegetables out the back and flowers and shrubs at the front with neatly edged borders and a lawn.[1] The nature of gardens was different though: the absolute centrality of home food production was being replaced by ornamental or low-care gardens and outdoor living spaces. Garden guides reflected this shift. *The Illustrated Planting Guide* (c. 1960) included 119 pages of trees and shrubs for home gardeners to choose from and information on making a pebble garden and growing bonsais – and devoted just five pages to growing fruit and herbs.[2] The gardening focus of the *New Zealand Woman's Weekly* in the late 1960s was on flower-growing and the preparation of indoor flower displays.[3] An analysis of census data shows that the interest in home vegetable production fell by 10 per cent between 1956 and 1971.[4] With that came a gradual loss of common gardening knowledge.

A changing focus

The description of garden features as a selling point had been steadily waning in real-estate advertisements since the 1930s. In the Christchurch *Press* for the first three weeks of January 1965, gardens or garden elements figured in only 17 per cent of advertisements for properties for sale or rental – the lowest of all samples since 1865. Concrete paths featured prominently, and just 4 per cent used phrases such as 'tidy section', evoking little more than a small shrub border and some lawn. Fruit trees, the most commonly mentioned element, appeared in fewer than 2.5 per cent of advertisements; trees and shrubs were next, and lawns trailed behind in less than 1 per cent. Where earlier advertisements had included detailed descriptions of fruit tree varieties, sections were now described as 'tidy', 'neat', 'easily maintained' or 'attractive'. The 'sun terrace' or 'patio', workshop space and, above all, shelter for the car became attractive selling points. Single or double garages and carports figured in 44 per cent of all advertisements, and these, along with fridges, 'dishmasters' and wall-to-wall carpets, now added perceived value to the proliferation of 'Summerhill stone bungalows with outdoor seating and within easy reach of the shops'.[5]

For some families, to be sure, little changed. William Noonan kept a large vegetable garden in Racecourse Road in Christchurch from 1965. His daughter Coralie, born in 1952, believed simply that 'all fathers had a vegetable garden in those days'.[6] Others made a concerted effort to learn the necessary skills – with varying degrees of success. Douglas Winter was born in 1913 to parents who showed no interest in growing vegetables. He married Olwyn Wallis (daughter of Bert and Annie Wallis, whose passion for irises is mentioned in Chapter Three) in 1948, an event that apparently prompted him to learn to grow food. Until his death in the late 1980s Douglas maintained a productive vegetable garden and orchard – enough in the 1950s and 60s to feed four boys 'with a constant supply of vegetables'. According to his son Gareth, however, he was an 'enthusiastic but hopelessly organised vegetable gardener, with great flushes of crops, and a garden that looked like it had been planted by someone on acid'.[7] Douglas 'battled' the intransigent couch grass and seldom applied fertiliser. At the rear of his quarter-acre section he had an orchard 'where apples and peaches struggled and failed to thrive, plums excelled, and where gooseberries and currants ... were grown in patchy rows'.[8]

Olwyn's story was different altogether. She traded plants with relatives and neighbours and created beds of bulbs and perennials:

> rose bushes [were] dotted around the lawns ... Muscari in large
> drifts, old hybrid freesias, innumerable gladioli ... large Opuntias in
> a raised box ... some wonderful Hedychium, growing in a damp but
> frost-free area underneath a window, and a large *Lonicera fragran-
> tissimum*, the shrubby winter-flowering honeysuckle.[9]

Douglas and Olwyn's boys were practically raised in this garden, and Gareth became a skilled gardener in his own right. He is an excellent example of how that economy of gardening knowledge – and plant materials – can be kept alive.

Modernity and leisure

By 1965 New Zealand had entered the age of modernism, and this extended to attitudes to gardens. The South Island's first 'garden centre', an idea imported from America that sought to encourage 'drive-in ... self-service' customers, was established in Christchurch in 1959.[10] Catering mainly to people interested in trouble-free and low-skill gardening, the centres seldom offered much in the way of expert advice and likely contributed to the gradual erosion of horticultural skills in the general populace. Gareth Winter, who 'haunted' a local garden centre in the 1970s until the owner eventually offered him a job, soon found he knew more than the other staff.[11]

New garden designs complemented the simplicity of garden centre offerings. *The Illustrated Planting Guide*'s instructions for making a pebble garden are revealing (and of particular interest to those seeking a Japanese influence in New Zealand's gardening past): 'Pebble Gardens, a modification in the styles of Oriental Gardens of a certain type to suit Occidental tastes and requirements, had their origins in Japan probably about 800 to 1000 years ago.'[12] Whereas the Japanese tradition was designed to encourage 'placid meditation', the modern version was 'most used for filling smaller corners, narrow betwixt-houses, and so on'. A pebble garden could be incorporated into existing garden forms and was simple to make: a layer of polythene beneath the stones eliminated the need for weeding, sparse planting made maintenance easy, and the garden could remain in place 'for several years at a time'.[13] The guide recommended the use of coloured scoria or marble to 'make some impact'. In Masterton, at least, this style became a trend: Gareth describes scoria being sold 'by the truckload' there as late as 1979.[14]

All of this was part of a shift that saw outdoor areas become an extension of the home's living and dining spaces. Matthews described it in 1960: 'the grounds become outdoor living-rooms … the house and garden merge into each other; there is no distinct line of demarcation'.[15] The new fashion can also be seen in McPherson's *Complete New Zealand Gardener*, revised in 1964 to include, among other things, a section on patios: 'Today the trend is to eat and entertain outdoors whenever weather permits' – and a patio was just the ticket.[16] Brian Morris (son of John and Florence mentioned in Chapter Six) and his wife Lois (daughter of Herbert Tyrell mentioned in Chapter Five) embraced the trend. In the 1960s in Invercargill Brian built a seating area surrounded by decorative gardens so they could relax and entertain outside.[17]

The home garden was gradually becoming a place of leisure rather than of necessary food production. Lawns bordered by shrubs and bedding plants began to replace the ubiquitous orchards and vegetable gardens of earlier times. Of the seven garden-centre advertisements in

Lois Morris with her son John in their tidy and decorative front garden in Invercargill, 1966.

Photo courtesy of Lois Morris

the *Press* for one Saturday in November 1976, five featured shrubs of various kinds, four specifically featured 'bedding plants', three mentioned herbs and only two offered vegetables. Among the decorative plants singled out were begonias, gerberas, jasmine, petunias, fire bush, azaleas, marigolds, carnations and a waratah.[18] South African shrubs were 'very much in vogue'.[19] Gareth Winter reported that in Masterton, potted canna lilies sold 'almost as soon as they arrived', native cultivars with bright foliage were in demand and deciduous trees were popular: 'each year we would sell hundreds of silver birches and golden elms as well as a good range of oaks, limes, ashes and elms. We sold more flowering cherries … than I care to remember.'[20]

The focus of older gardens was changing as well, although many maintained established features of the past. The 1980s vegetable garden at the homestead of Waihi Bush, now home to Max Musgrave's son, was substantially smaller than it had been in Max's day, but productive fruit and nut trees were retained along with large rhododendrons, a massive copper beech, gums and a huge pine. The children browsed in the raspberry patch, and a number of new blackcurrant bushes were planted

(Ethel) Margot Musgrave with harvest, Waihi Bush, 1973.
Photo courtesy of David Musgrave

Mark Wanhalla by his family's vegetable garden, Christchurch, 1982.
Photo courtesy of Angela Wanhalla

alongside the old Roaring Lion gooseberry bushes in the cherry house.[21]

Many householders still believed it was important to keep the garden tidy. Dallas Matoe remembered that their garden 'was tidy because of my stepfather's personality. He was quite meticulous and thorough.'[22] Stan Wanhalla's 1980s garden was 'immaculate; the lawn neatly clipped and edged, the vegetables always in straight lines, the garden beds always weed-free'.[23]

As time wore on another backyard feature complementing patios, barbeque areas and other 'Californian' features became commonplace: the swimming pool. In Auckland alone, by 1984 there were 8400 private swimming pools.[24] Even these could be claimed and redesigned by environmentalists. The Permaculture Institute of New Zealand published an article in 2000 about an unchlorinated swimming pool that utilised a Rudolf Steiner-inspired 'flow form' and wetland settling area.[25] Apparently, a number of people in New Zealand were interested in such innovations; the organisation listed 32 'bioregional' contacts around the country.[26]

The transmission of garden knowledge

As the workload in gardens decreased, parental expectations of assistance from children also diminished and the golden opportunity to learn by example was eroded. Although Lachy Paterson was enlisted to mow the lawn in his father's abundant 1960s Whakatāne garden and occasionally helped his grandparents to bottle tomatoes and beans from their enormous vegetable plot, he did not learn how to cultivate plants and tend the soil.[27] Apart from mowing lawns and growing experimental patches of radishes or sunflowers, Brian and Lois Morris did not expect their children to help in their 1200m^2 garden in the suburb of St Martins. And Christopher Musgrave, whose parents David and Diana took over the old family farm at Waihi Bush, had little to do with maintenance of the productive home garden: 'I don't remember my parents ever really teaching me how to do anything [in the garden].'[28]

Not all families enjoyed a stable home with an established garden in which to experiment. Raised in a family still reeling from the impact of colonisation in the nineteenth century, Piki Te Ora left his Hāwera home to escape his father's alcoholism, an addiction he later developed himself. Piki's son Dennis Matoe also absconded at an early age. He married Leigh Weymouth, who came from a similarly challenging background, but abandoned his young family in the early 1970s shortly after his own son Dallas was born. Dallas's stepfather maintained a garden at their Philipstown home where they lived from about 1977. The property had a number of fruit trees, roses and a small vegetable plot. Dallas recalled, 'He used to make me mow the lawns and stuff, turn over the garden … But I sort of resented him and I'd close off. If I knew then what I know now I would have really enjoyed it. But … I tried to get out of there as fast as I could.'[29] There was little opportunity to share gardening skills effectively in such an environment: 'If you're living tribally you're living in a larger community that has a skill and knowledge base, like with gardening,' Dallas remarked in 2010. 'Then you get good mentoring as well …'[30]

Gardening struggles of the young: 1970s and 1980s

Despite the influences of modernism, many people still maintained good vegetable gardens, and for those with the desire to grow food there were

Sylvia Paterson with the family's impressive vegetable garden, Whakatāne, 1984.
Photo courtesy of Lachy Paterson

plenty of options. A catalogue from Webling and Stewart of Wellington published in 1973 contained 54 pages of vegetable seeds, including 14 different pumpkins and squash.[31] A 78-page seed catalogue from Yates (c. 1984) had 53 pages of vegetable seeds to choose from.[32] But many young people struggled with gardening.

In 1970 Coralie Noonan married Stan Wanhalla, son of William and Waitai of Taumutu at Te Waihora. Prior to European settlement Te Waihora had been a thriving community, and was the southernmost point where kūmara could be grown. Stan could grow a good crop of potatoes but had little experience of producing other vegetables. Marriage spurred him into action, as it had Douglas Winter. At their flat in Islington, Christchurch, Stan planted silverbeet and tomatoes, and he quizzed his workmates endlessly about gardening in an effort to learn more. At the landlady's home across the road he experimented with a lawn, a chicken coop, and a vegetable garden that supplied Stan and Coralie, the landlady and three of Stan's brothers.[33] By 1978 the couple owned a 'tiny' section in a new subdivision in Hornby where they established a garden from scratch: they trained a grapevine along the back fence, and

Angela Wanhalla by her father Stan's vegetables, Christchurch, c. 1982.

Photo courtesy of Angela Wanhalla

there were rows of potatoes, a currant bush and copious amounts of silverbeet. 'I always remember corn as well: rows and rows of corn,' their daughter recalled.[34] Stan's experimentation had paid off.

As a young man in the late 1970s, Lachy Paterson shifted south to Palmerston, Otago, where he stayed with a 'hippie' friend who had a big vegetable garden and chickens. After moving to an old farmhouse in the early 1980s, Lachy tried his hand at making his own vegetable garden for the first time, but despite a lot of digging it was generally a failure: everything went to seed and became weed-infested. Only when he teamed up with his first wife, who was a more experienced gardener, did he begin to hone his horticultural skills.[35]

A similar example comes from the edge of the Waitākere Ranges in the mid-1970s, where Christine Nash grew a few vegetables and flowers. Despite the fact that her mother Joan had been in the Land Army

Cuppa tea time, St Albans Gardening Group, c. 2009. Christopher Musgrave is second from right.

Author photo

in Britain during World War Two, no gardening knowledge had been passed on to Christine, and she herself passed little on to her own children. Her sons Jared and Jolyon White both became gardeners in time: Jared learned horticulture at Lincoln University, where he was inspired into organics by Bob Crowder (see below). Jolyon, on the other hand, was living in Golden Bay in the 90s when a friend kindly put in a large vegetable garden for him. Without the necessary knowledge and support the garden overwhelmed Jolyon and nothing survived.[36]

Christopher Musgrave also struggled to start a garden. In 2009 in an effort to learn from scratch he joined the Soil & Health Association and the St Albans Gardening Group.[37]

Some people picked up horticultural skills in less orthodox ways – through growing drugs.[38] Many gardeners have cultivated marijuana, opium poppies and San Pedro cactus in their home gardens in this country, but by and large they are reluctant to disclose their growing methods.

New ideas: Garden practices in the 1980s

Compost bins were a common feature in many gardens throughout the 1970s and 80s, but making compost involved a lot of digging and turning. Those with a bit of money and a desire to ease the work-

A beer on the back lawn after gardening, Whakatāne, 1987. Left to right: Glen Paterson, his father Gavin and Gavin's father Len. Note the runner beans and plastic compost bin in the background.

Photo courtesy of Lachy Paterson

load welcomed the arrival of the compost tumbler, which began attracting attention from the late 1970s. One brand, the Sutton Compostumbler, was introduced to New Zealand in 1977 as 'a revolutionary new labour-saving means of returning to the soil organic wastes – so simple – so convenient!'[39] A review of the device by a member of the Soil Association's Canterbury branch was enthusiastic: 'It certainly will encourage people to make compost, particularly when the spirit is willing but the flesh is weak.'[40] One Auckland inventor went so far as to construct a wind-powered rotating compost unit.[41]

Another concept that captured gardeners' imagination, particularly from the 80s, may also have been driven in part by a desire to save time. Companion planting, 'the planting of different crops in proximity for … pest control, pollination, providing habitat for beneficial creatures, maximising use of space, and to otherwise increase crop productivity', blurred the boundaries between the vegetable garden and flowerbeds that until then had been sharply drawn.[42] To repel pests, 'Aromatics such as marjoram, lavender, santolina and lemon balm should be scattered informally among crops,' one proponent wrote. 'Marigolds put off a substance that keeps nematodes at bay. The roots of both oats and flax excrete compounds that inhibit the growth of some harmful soil fungi.[43] Although the various plant combinations could be bewildering, *Soil and Health* suggested there was no need to panic. Research into plant needs was beneficial, and anyway, if one gardened 'with love and joy' then 'one can't go wrong'.[44]

Companion planting harked back to an era of so-called 'cottage gardening', and was linked with a notion that probably caused offence to some: 'weeds' were simply plants in the wrong place. Tim Maples suggested that it was ecologically advantageous to leave rough, untamed areas in the garden as habitat for beneficial fauna. 'Clean cultivation is an accepted practice, as a tidy garden is assumed to be a healthy one':

> Although weeds do harbour pests and can give crops competition, beneficials depend on them for both shelter and bugs to prey on. As with any means of control, weed-pulling should not be exercised arbitrarily, and only after some deliberation on your part.[45]

The presence of weeds in a garden could even be cause for celebration, according to the Soil Association at least:

> Once you come to terms with the fact that every garden will always have weeds if the soil is at all fertile, and that there is a great deal to be learned from observing them, the problem can be approached in quite a different and indeed, in a much more positive way.[46]

This radical departure from customary garden advice constituted a new salvo in the debate over chemicals and intersected neatly with the growing desire for low-maintenance gardens.

An organic vision

Organic gardening began to come into its own during the late 1970s and 80s. Bob Crowder, who had established an organic gardening and farming research and demonstration area at Lincoln College (now Lincoln University) from 1976, was part of this picture. His mission was to propound an organic vision for New Zealand and inject some 'greater scientific credibility' into the debate. His teachings were circulated widely both at home and abroad.[47] Bob drew inspiration from organic research institutions such as Rodale Institute in the United States, the Research Institute of Organic Agriculture in Oberwil, Switzerland, and

Bob Crowder on the cover of Soil and Health *magazine, Summer 1988.*
Soil & Health Association of New Zealand

Witzenhausen at the University of Kassel in Germany.[48] The 1994 conference of the International Federation of Organic Agriculture Movements (IFOAM), which Bob organised at Lincoln University, drew hundreds of delegates and a supportive message from Prince Charles, and left an indelible imprint on the organic movement in Aotearoa.[49]

The Henry Doubleday Research Association of New Zealand (related to the British organisation of the same name, founded in 1954 by Lawrence Hills in honour of Henry Doubleday's contribution to research into the properties of comfrey) commenced operation in 1977. The association promoted research into 'basic and improved methods of organic farming and gardening' and, in particular, into 'the use of Russian Comfrey and other plants' for agricultural and horticultural uses.[50] The association's first newsletter put the research aims into perspective: 'Many people today are beginning to realise how much our communities are becoming dependent on artificial aids to everyday living. We have artificial environments, food, entertainment and so it goes on.' After hailing the advent of the environmental movement decrying this artificiality, the chair of the organisation, Dave Woods, described how 'indications of dissatisfaction are shown by the number of persons who are now going back to the land and forming "homestead" or "commune" styles of living, or even just working on their own eighth-acre plot':

> It has been said many times that oil is going to run out, but so are many other less obvious items such as fertile land for growing our food. The fertile land will 'run out' because of urban sprawl and because of abuse, or inefficient use by those who do not know any other way.[51]

The sentiment, so representative of the new environmental movement, is strikingly similar to that expressed in the 1940s by the composting movement, Forest & Bird and some Christian churches.

The Soil Association reported events at no less than 14 branches around the country in 1979, and 17 in 1986.[52] All over the country, it seemed, people were quietly adopting and promoting organic gardening methods. George and Marjory Maslin first learned about organic gardening when they worked at a 'native school' in the late 1940s in Mōkai, a rural Māori community in the North Island. The couple moved to Christchurch in 1951 and created an abundant organic garden there. As George wrote in a 1984 essay, he viewed his efforts as 'one of the most important things a Christian can do: set a good example of soil use in your own back yard'. In this essay – essentially a guide to social reform – George argued strongly for an environmental approach and equitable distribution of the 'earth's bounty'.[53]

The quiet decline of the large garden

As the nature of gardens was changing, so too were cityscapes. With expansion, increased traffic, higher-density housing and more high-

George and Marjory Maslin's back garden. The rotary clothesline stands amid a verdant compost-fed vegetable garden. Christchurch, c. 1989.

Photo courtesy of George Maslin

Marjory Maslin is dwarfed by her corn, Christchurch, c. 1989.

Photo courtesy of George Maslin

rise buildings, the call to preserve urban green space became more urgent. In 1975 Christchurch City Council's assistant town planner drew attention to the destruction taking place in the name of development: 'Developers have tended to clear sites completely, including the removal of all existing vegetation ... and mature trees which have taken many years to grow have been destroyed in a matter of minutes.' The council discussed a system to protect mature trees, a suggestion that at first drew criticism from the *Press* in an article headed 'Tree scheme seen as madness'.[54]

All over the country, large urban properties with established gardens were being subdivided to make room for infill housing. Subdivision destroyed old gardens and with them the gardening culture. William and Minnie Basset's plentiful garden in North Parade, Christchurch (see Chapter Three), was subdivided in the 1980s to make way for two houses surrounded by a desert of concrete; Iris Keenan's wonderful Hastings garden, on which she had lavished so much love and attention (see Chapter Six), was destroyed during the 90s when the family replaced the old house with two flats.[55] These were but two gardens among many lost to subdivision.

Of course, smaller sections meant less space for gardening. *The Ultimate New Zealand Gardening Book* (1994) discussed the concept of inner-city living and encouraged those with small properties to consider turning their outdoor area into a patio.[56] Acknowledging the constraints of smaller spaces, Diana Anthony's *Creative Sustainable Gardening in New Zealand* (2000) contained an entire chapter on 'no soil' and 'small-space' gardens.[57] Not only was knowledge about gardening fading; the gardens themselves were disappearing.

The problem of urban intensification and loss of gardens and gardening knowledge worsened as the century wore on. Christchurch resident Peggy Kelly felt this keenly. An in-filled neighbourhood with ever-taller fences was, in her estimation, a blow to 'the greatest good' – beauty: 'My nature is nurtured by beauty, even if it's the way a swallow flies, how the light falls on the pine needles at sunset. The magpies shouting at each other are beautiful ... These six-foot fences around tiny courtyards ... are dead ugly.'[58] Intensification, and gaps in the city plan, meant notable trees could be cut down by developers, resulting in the death of habitat and a painful loss of beauty. During the 1990s, as her suburb filled up with units crammed onto ever-smaller sections and green spaces disappeared, she and others around her bore the brunt of what was, in her eyes, a spiritual and social crisis. Wellington community worker Charles Barrie echoed this thought in 2009: 'If we cut down all our trees, we won't just die physically but also spiritually, which will be much more painful.'[59]

Urban infill housing was not the only cause for this, however; the decline in gardening was visible in smaller towns too. A 2007 study of 120 gardens in Lincoln township, Canterbury, ascertained that a typical section comprised 40 per cent lawn and 20–30 per cent paved surfaces. 'Woodland was rare, shrubberies were common'; there was a 'greater prevalence of herbaceous borders in front yards than in back yards. Vegetable gardens were not well represented.'[60] Where residents had space to garden they mostly kept just a lawn and shrub borders. Some owners of newer homes felt little personal connection with their garden, saying simply that the developer had created it for them. It was a far cry from the gardens of even 50 years before.

The late 1980s saw an 'upswing of interest' in flower gardening. Gareth Winter described a 'swing back to faux "cottage gardens", with a corresponding focus on perennials and annuals and a slow decline of interest in vegetable gardening'.[61] More and more the mass market for plants was changing to meet the needs of those with space for a few elegant pots in their courtyard and scant knowledge of or desire for gardening. In her history of New Zealand gardening, Bee Dawson has noted that although the move to large garden centres 'initially seemed to promise more efficient plant shopping, both the range of plants and the art of gardening quietly declined … New releases and traditional favourites were there in their hundreds; connoisseurs' delights were not.'[62] Dawson is probably correct to suggest that this renewed interest was prompted by the 1987 stock market crash. It is not surprising, therefore, that the end of the 1990s was also 'the end of the golden spell for nurseries': in 1997 Gareth Winter sold his Masterton business, and many others also closed around that time.[63]

The growing world of community gardens

Examples abound of community gardens in New Zealand prior to the 1990s. Collective inner-city plots appeared from time to time: during

World War Two the Women's Land Army collaborated with the Compost Club to make a vegetable garden at Abberly Park in Christchurch; the Sunlight League did the same.[64] Some communal gardening was taking place in Christchurch's Avon Loop during the 1970s, and the Peterborough Community, a group of 'six households in a row with joined-up back yards', produced vegetables and fruit, some of which were sold at Piko Wholefoods store.[65] The urban communities Chippenham and Mansfield in Christchurch were linked to a rural commune in Oxford called Gricklegrass, where residents maintained a communal orchard and flower and vegetable gardens.

Communal living and gardening were viewed with curiosity and some scepticism by the mainstream. Those working in the Peterborough gardens were referred to by one newspaper as 'an anarchist group', and at Gricklegrass the *Press* described 'bare-breasted girls driving tractors through fields of organic barley, [and] upwards of 30 youngsters dropping out of the consumerist rat-race'.[66] Ami Kennedy (daughter of Tom and Ali Kennedy mentioned in Chapter Six), who was born at Gricklegrass in 1974 and lived there until 1980, recalled the community's philosophy of 'sharing resources and working for the community'.[67]

The 1980s saw a thriving community garden take root in Aro Valley in Wellington, the produce from which was shared through a communal food distribution centre.[68] In Auckland, thanks to the initiative of Paul Lagerstedt, the Kelmarna Community Gardens at Herne Bay were established in 1981, and in 1983 Belmont Cooperative Workers' Trust set up a garden, also with Paul's help in conjunction with Jenny Bolton.[69] The Belmont garden boasted 'French-style raised beds filled with … organic material and covered with a thick layer of Devonport Borough Compost'. No 'artificial, chemical fertilisers' or 'poison sprays' were used.[70] The Golden Bay Community Gardens were established in 1986 following a community composting trial in Tākaka in 1985.[71]

The reforms begun under the fourth Labour government from 1984 gradually made the need for community gardening more overt as the challenge for many to 'make ends meet' became steadily more difficult. In the first three months of 1991 the demand for Salvation Army food parcels rose by 76 per cent; by 1993 there was a tenfold increase.[72] In Lower Hutt a single food bank handed out 4400 food parcels in the year to June 1994, up from 640 in 1989.[73] But the problem was not simply a lack of money. A

Kelmarna Community Gardens, Auckland, 2010. Note the use of raised beds, and plastic bottles to protect plants.

Author photo

Devonport Community Gardens, Auckland, 2010.

Author photo

certain individualism had crept in, leaving people more vulnerable in difficult times, as Anglican bishop Whakahuihui Vercoe wrote in 1994:

> We talk about being poor today, but I grew up [during the Great Depression] in a poor society with no money and no work. Everybody was unemployed. But people worked to sustain themselves, to grow their own food and to buy only the bare necessities of life ... Our society planted for the whole of society, not just for individual families. A percentage of your vegetables ... was designated for the common good of the hapu.[74]

As government reform programmes brought in tougher measures, genuine hardship was increasingly felt in those communities with the fewest resources at their disposal. Community gardens were among a raft of community initiatives that began to evolve, with varying degrees of success.

Te Whare Roimata was founded in Christchurch's inner city in 1986 with exactly this in mind. Ophelia Swarez-Chambers, Tony Church and Willie Kong were instrumental in developing community gardens with the Anglican City Mission to grow food for distribution and sale in a

Willie Kong in the Old Vicarage community garden, Governors Bay, c. 1999.
Author photo

café. A Neighbourhood Gardens Project, set up in 1989 'as a self-help response to unemployment and the lack of affordable vegetables', initially utilised five inner-city private gardens, to which three adjoining gardens were added in 1997.[75] Another site, this one on council land, was 'leased' to Te Whare Roimata in 2001. Te Whare divided this land – 9000m² in extent – into individual allotments for residents and groups, reserving an area for its own use. By early 2002, seven community groups were involved in this project.[76]

Porirua's Te Maara @ Cornwall community garden was established in the late 1990s in a state-housing area. Initially set up by members of St Anne's Church, the garden would 'give people a sense of community, of belonging … a sense of self-worth; of being needed and valued', as well as teaching the principles of organics and permaculture and 'providing those in need with freshly grown vegetables' through the church's food bank.[77] The garden had fallen into disuse when John Poppleton, a member of Keep Porirua Beautiful, recognised an 'opportunity for the residents … and passionate gardeners … to get together and share gardening knowledge'.[78] John drew in schools, Titahi Bay Horticultural Society, Wellington's Sustainability Trust and locals: in all, over a thousand people had participated in the project by 2009.[79] The hoped-for result – full ownership of the project by the community – necessitated what one community worker called a 'resiliency towards failure': 'it takes time to learn how to grow things in a way that won't fail'.[80]

Some communal gardens arose from protest action, such as at Bastion Point in 1977–78, where Māori activists and supporters occupied a 24-hectare site in an attempt to prevent the subdivision of Ngāti Whātua's ancestral lands. The occupiers planted extensive food crops to support themselves. Sadly, these were destroyed by army trucks when the protesters were forcibly evicted.[81]

Community gardeners in Auckland's Basque Park were also ousted in 2002. Fortunately for this group, St Columba Church offered land to the main organisers; the resultant Grey Lynn Community Gardens now provide space for a variety of community activists, anarchists and environmentalists to bring their ideas of community interaction, support and ecological stewardship into the world.

In Christchurch the Edgeware Community Gardens were at the centre of a tempestuous campaign to preserve inner-city land from intensive

Community gardening leaders Peggy Kelly and Bill Sykes 'turn the sod' in the Edgeware Community Garden, Christchurch, c. 2009.

Author photo

housing development from 2007.[82] After the council demolished a 70-year-old community-built swimming pool on the site to make way for a subdivision, local residents established a garden there to demonstrate the value of the land. A council working party set up to explore options for the site called for submissions, over 12 per cent of which suggested that expanded community gardens or allotments would be a tremendous local asset.[83] Some cited the loss of garden space as a real concern that needed to be addressed. As one person wrote, 'Some flats/units have no garden to speak of … To have a patch of land to create something beautiful … can be a wonderful and empowering thing.'[84]

Massey Community Garden in Waitākere started in 2009 as a way for a group of flatmates 'to get to know the neighbours'. The friends organised 'Big Digs', where residents in the street got together to garden in the park.[85] The community garden in Wellington's Aro Valley, according to community worker Charles Barrie, was merely a successor to the communal gardens of the 1980s and, further, to the pre-European Te Āti Awa kāinga there. Charles believed it was 'crucial' for people 'to learn to live in the community again in a way that's quite safe. We've moved into individualistic ways where people don't know their neighbours.' He saw great value in sharing: 'We need to stop stamping "this is mine" and "this is yours" onto everything … [and] community gardens are an amazing way to test the parameters on that.'[86]

People felt a sense of achievement in these gardens. As one Christchurch community gardener said, contributors gained 'a sense of pride from building planter boxes, making compost, germinating seedlings and watching their plants grow'.[87] As food prices escalated, many people were finding this a way to reconnect with a culture of productive home gardening that, by the turn of the millennium, had all but disappeared.

A Community & Home Gardens group formed in 1998 in Christchurch under the umbrella of the Organic Garden City Trust, about which more shortly. Its role was to relieve 'poverty in our community. That is, poverty of resources, poverty of learning, poverty of empowerment and poverty of health (wellness).'[88] In 2000 the group became the Christchurch Community Gardens Association with 10 member garden groups listed.[89] They noted a dramatic increase in community gardening from 2008: in mid-2009 there were around 20 community gardens in Christchurch, and in 2019 the number had risen to 24.[90] In Auckland the number of community gardens also increased, from 20 in 2009 to 73 in 2018.[91] Nationwide there were at least 152 community gardens in that year.[92] The network of community gardeners flourished and groups supported one another in their work. Vandalism in one garden would be met with generosity from others, as when in 2018 Richmond Community Garden, New Brighton Community Garden and Ōtākaro Orchard gardeners between them donated 200 plants to Wainoni Avonside Community Garden after it was targeted by thieves.[93]

Novice community gardeners picked up useful tips from experienced home gardeners. Margaret Jones, the doyenne of composting in New Zealand, at 89 was still taking groups from the Crippled Children's Society (now CCS Disability Action) community garden around her own extraordinary 800m² section and sharing her vast knowledge. Margaret grew 52 fruit trees in her Waitākere City garden, among them cherimoyas, tamarillos, pomegranates, persimmons, apricots, lemons and avocados, and passing school children were encouraged to help themselves from the hedge of feijoas, guavas and mountain paw paws along the front of her section. A herb garden contained over 30 varieties, many of which she used for medicinal purposes. Margaret was named Auck-

Margaret Jones in her Henderson garden, 2002.
Photo courtesy of Margaret Jones

land Gardener of the Year in 2008. When told the news by *New Zealand Gardener* she replied simply, 'No! My garden is shit!' 'Yes,' they replied enthusiastically, 'we know about the three "s" rule: seaweed, sawdust and shit!' Margaret appreciated the honour, and explained her approach: 'My garden is a practical home garden. It's not there to be beautiful.'[94]

Green cities and urban food security

Growing public discussion about the risks and extent of chemical use began to have an impact. In 1997 Auckland City Council reported that 1413 residential properties were listed on their 'no-spray register' – a list of residents who preferred to maintain their frontage rather than have the council apply herbicide.[95] Various moves were afoot around the country, many with an organic focus, to improve the capacity for cities to feed themselves. The Organic Garden City Trust in Christchurch was one initiative. It began partly with the vision of Mayor Vicki Buck, who suggested at the 1994 IFOAM conference that Christchurch – already a Garden City – could become the world's first *organic* Garden City.[96] A 1997 workshop resulted in the formation of a legal entity to coordinate and support efforts to move

Christchurch in this direction.[97] Rod Donald, co-leader of the Green Party and former member of the Peterborough Community, was one of the drivers behind the creation of the trust, which brought together people involved in community, home and school gardens as well as commercial producers and retailers.[98]

Another initiative called Kids' Edible Gardens introduced organic gardening techniques to primary school children, who were encouraged to share that knowledge with their parents and whānau. Driven by Lily White, Rod Donald and Frankie Dean, the project aimed 'to re-establish that link between growing, and where ... food comes from'.[99] Whereas many children a few decades earlier had grown up watching and helping their parents to grow food, by 1997 some simply had no idea that vegetables grew in the ground. According to Ami Kennedy, now working for the initiative, the project was 'giving children a connection to where their food was coming from', enabling them 'to have a positive impact on their environment'.[100] By 2001 Kids' Edible Gardens had reached over 2500 children.[101]

By 2009 several such programmes were on offer. In one, a Kakanui gardener delivered heritage potatoes to 23 schools in Otago as part of

Children learn to make compost as part of the Kids' Edible Gardens initiative, Christchurch, 2008.

Photo courtesy of Lily White

a nationwide 'spud in a bucket' programme that included over 5000 children.[102] In Auckland the Garden to Table Trust, established in 2008, worked with three pilot schools 'to allow children … to enthusiastically get their hands dirty and learn how to grow, harvest, prepare and share fresh, seasonal food'.[103] The community agency Te Ora Hou Ōtautahi developed a community garden at their site in Christchurch in 2008 in order that young people might gain 'knowledge of how to provide basic food … for the family' and 'reconnect with the land, thereby gaining sense of meaning and purpose'.[104]

When we consider the Matoe family history running through this book, from Turi and his kūmara gardens at Pātea, through land losses in the nineteenth century to the disintegration of family networks that supported the transfer of gardening know-how between generations, the value of school-based gardening schemes makes good sense. In 2010 Daisy Matoe, daughter of Dallas Matoe and Sarah Butterfield, was part of a gardening club at her primary school, learning about vegetable growing and enjoying the experience.[105]

Transition Towns and other initiatives

In 2007 a new idea emerged that dovetailed neatly with this 'green city' thinking. Transition Towns is a 'vibrant, international grassroots movement that brings people together to explore how we – as communities – can respond to the environmental, economic and social challenges arising from climate change, resource depletion and an economy based on growth'.[106] The initiative provided a platform for sharing skills and strengthening communities, and spread rapidly as groups keen to take ownership of their environment and local economy formed around the country.

The first was on Waiheke Island, where an OOOOBY (Out of Our Own Backyards – a food exchange scheme for produce grown in home gardens) began in 2007 alongside a community-supported agriculture project and 'the Fabulous Fruit Tree Initiative', which aimed to establish 20,000 fruit and nut trees on the island.[107] Sustaining Hawke's Bay Trust launched Transition Hawkes Bay in 2007 and established a 'food group' with a community garden and seedbank among its initiatives, and north of Dunedin the Waitati Edible Gardeners Group ('the Weggies') was set up. Along with garden tours, workshops and school garden beds, in 2009 the

Helen Ross helps with Diane Shannon's compost as part of a St Albans Gardening Group working bee. Christchurch, c. 2009.

Author photo

Weggies negotiated the establishment of an allotment-style community garden with individual plots.[108] In Christchurch the St Albans Gardening Group began holding working bees where members assisted one another to simplify food production in their gardens, and held regular potluck dinners to enable group building and information sharing.[109] Transition Timaru, formed in 2009, held a fruit tree-pruning workshop as one of its first projects; and in Ōamaru, local advocates organised a workshop on creating 'urban Edens', or edible backyard gardens.[110]

In the winter of 2007 the Garden Development Project got under way in Dunedin. Members of St John's Anglican Church and friends helped 'families struggling with their food budget' to establish vegetable gardens and compost systems. Each family was allocated a gardening mentor to provide ongoing support for a year and was encouraged to help in another family's garden nearby.[111] Instigator Jolyon White wrote:

> I forget so often that the way the world works is not inevitable. Things have not always been as they are, and will not stay so ... it seems whenever people talk about pressing environmental issues: plastic bags, ocean pollution, greenhouse gases, food prices, peak oil ... the solutions discussed resemble a way of life and skills my grandparents' generation took for granted.[112]

Like Charles Barrie, Jolyon's view on gardening was inspirational:

> I believe gardening is part of re-engaging our communal imagination ... I can imagine backyard fences coming down. I can imagine gardening becoming as widespread as it was when my grandmother was a child ... I'd like to see city roundabouts growing lemons, apples, rhubarb and silver beet. I'd like to live in the type of community that has time to take a dozen zucchini next door because you can't eat them all.[113]

Following the Canterbury earthquakes of 2010–11, the term 'food resilience' became commonplace in New Zealand. Golden Bay group Earthcare Education embarked on a nationwide 'Localising Food' tour in 2012–13:

> to empower communities in local food resilience and create healthy, enriching local food cultures throughout NZ. This mission came about as a response to the rapid decline in food security, escalating costs of food, and the increase in natural disasters, especially earthquakes and floods which have devastated some bioregions in New Zealand.[114]

Earthcare Education assisted communities to write 'food resilience plans', and the tour resulted in 'many local food networks', with 'home growers upskilled and properties transformed into food havens ... Thousands of home-saved heirloom seeds were exchanged, school seedbanks formed and food gardens established.'[115] A film produced from this tour, *Edible Paradise*, described how a food commons vision was being realised in New Zealand.[116]

The concept of the food forest – a planted space that uses food-producing perennials to mimic the 'layers' of a natural forest ecosystem – also began to receive attention. After a few years the 'forest' becomes mostly self-managing and is therefore more resilient than a conventional orchard, which requires ongoing maintenance.[117] Food forests are 'a way to build diversity, resilience and security back into our food system'; they recreate the idea of commons – spaces for gleaning from – and challenge a paradigm centred on private land ownership.[118] What was likely the first food forest in New Zealand was established by Robert and Robyn Guyton in Riverton in the early 1990s. This started with the establishment of native trees but evolved into a food forest, in part because Robyn wanted to recreate the kind of rambling garden her grandmother had. They were influenced by Masanobu Fukuoka's *The One-Straw Revolution* (1975) – which set out the tenets of 'do-nothing' or 'natural farming' – and Bill Mollison, co-founder of Permaculture. Not that they have 'done nothing': together the Guytons have grafted and donated some 10,000 heritage fruit trees throughout Southland. Robert's goal was to allow their garden to speak – 'it is a being on its own' – and to minimise their management of the forest in order to learn from it.[119] Their influence has been significant. By 2017 an estimated 250 food forests had been established in New Zealand, most planted within the last few years and many on public land.

Gardening for an authentic connection to the land

Home food gardening was on the rise again, and garden centres reported unprecedented sales of fruit trees and vegetable seedlings in 2008.[120] In 2009 Waimea Nurseries in Nelson stated that fruit-tree sales had doubled over the previous two years, and *New Zealand Gardener* published a special edition on growing fruit trees, 'the forgotten heroes of our gardening heritage'.[121] Along with the nostalgia value and concerns about issues of food security in a post-oil environment, this renewed interest in planting fruit trees reflected rising costs: fruit and vegetable prices rose by 12.5 per cent between October 2007 and October 2008 – even before the effects of the global recession had set in.[122] It was like a repeat of the early colonial environment in New Zealand when exorbitant food prices provoked settlers to get busy and produce their own.

It wasn't just food growing that was catching on, however. There was also a noticeable revival of interest in native plants. Richard Tankersley went to considerable lengths to learn about which natives to plant in his Addington garden in Christchurch. To his way of thinking, native plant species were like te reo Māori: too special to lose. They were also a useful addition to the ecosystem, and would contribute to the wider community efforts of groups like the Addington Bush Society (later the New Zealand Ecological Restoration Network) in reconstructing habitats for native birds and other fauna in city spaces.[123] One garden with a native focus displays a desire for authenticity: Ngā Tipu Whakaoranga o Tutekawa ('the plants that sustain us'), a garden at Motukarara near Banks Peninsula that opened in 2006, features koru-shaped raised beds planted in species of cultural significance to Canterbury Māori.[124]

Max Podstolski and his wife Clare Reilly, both Cashmere artists, sought to protect the local environment, and in particular native birdlife, through gardening with native plants, suggesting a deepening connection with the New Zealand landscape.[125] They saw themselves 'as indigenous artists'.[126] Jennifer Barrer, a Cashmere gardener and granddaughter of Guy and Grace Butler, mentioned in Chapter Four, viewed her relationship with the land in the most intimate – that is, spiritual – sense. 'I know the land. I understand the land. I understand the land in a way like an ancient spirit … I know the land in that kind of way, where you're so much part of it, in tune with it – all the nuances – that it's almost like *they* shape you, or what you're going to do.'[127] For Jennifer, the land had a transformative effect. As a fifth-generation New Zealander, she believed a deeper substratum of being existed that could be accessed; and the garden was an essential component of this process. '[A]s soon as I have my hands in the earth it really does ground me, and provides a certain kind of order of essential things.'[128] Jennifer's description of gardening as helping to link her, on a spiritual plane, to the whole landscape, to become part of it, echoes Charles Barrie's sentiments that 'all elements of life are part of one living being. Our consciousness and awareness is something that emerges from the landscape in which we

Max Podstolski and Clare Reilly's Cashmere Hills garden, Christchurch, 2009.
Photo: Clare Reilly

live.'[129] For Jennifer, as for Max and Clare, gardening helped give life to a whakapapa, a genealogy anchored in the physical landscape.

Marae gardens and Māori gardening

From the mid-1990s some marae showed renewed interest in creating collective gardens. With help from his whānau, leader Bill Solomon developed a number of gardens on Takahanga Marae in Kaikōura, an area once renowned for its kūmara gardens. Bill's sister-in-law Darcia Solomon helped develop a 'kaumātua garden', a place where families could plant something in remembrance of deceased elders. Bill himself created a massive vegetable garden, complete with compost heaps and worm farms, to feed the marae. The gardens were spray-free: 'We never know what the sprays ... consist of. I don't mind if the food has a slug in it,' Darcia confessed. Another garden became a bush-like sanctuary, a place to clear one's head. According to Darcia, the gardens 'were all about people': in creating them, Bill had brought the gardening traditions of his parents and grandparents onto the marae.[130]

On the east coast of the North Island, under the guidance of Roger White (descendant of the Whihongi whānau mentioned in Chapter

Darcia Solomon in the kaumātua garden at Takahanga Marae, Kaikōura, 2010.
Author photo

Vegetable gardens at Takahanga Marae, Kaikōura, 2010.
Author photo

Three), primary healthcare provider Ngāti Porou Hauora worked with communities and, in particular, with men and young boys through the Mana Tāne group, to establish organic community gardens. At Rangitukia they grew kamokamo, in Te Araroa kūmara and in Tiki-tiki Māori potatoes. Roger's motivation for the gardens was around encouraging physical activity and 'making food more accessible to our whānau'. Food was grown in traditional ways using the maramataka, the Māori calendar, and sustainable methods that ensured gardeners 'looked after the soil'. 'It's really rewarding for people … to have a good mixture of the older generation and younger … there's a lot of knowledge transference.' Old practices were shared: 'We do our best to grow [things] the way our ancestors grew them.'[131] One of the gardeners in the group had learned a method for sprouting kūmara tipu from a man in his eighties, who had learned it from his grandfather: 'He leaves them in muddy water for a couple of days and then he plants them out into the garden. It promotes strong roots and shoots. This makes the kūmara grow big and strong in the ground.' Potatoes were stored in a pākoro, a pit in which potatoes and ferns are layered then covered with soil.[132] Roger shared Darcia Solomon's sanguine view of pests: 'Sometimes a few pests get into our kai, but we just cut it out and eat the other half … Traditionally, our people didn't have chemicals, so that's the way we do it as well.'[133] The community gardens brought people together: 'When you look at traditional ways of living for Māori people, that's how it was. Everything was done together. That's something that's missing in today's society.'[134]

By the end of the first decade of the twenty-first century the concept of reviving old practices was taking firm root. Based on the knowledge he had learned as a child from his uncle Honiti Apiti of Kāwhia, and with the aim of 'preserving the stories and traditions, not only of gardening but also of food preservation', Wiremu Puke (Ngāti Wairere, Ngāti Porou) established Te Parapara, a 'traditional Māori garden' in Hamilton Gardens.[135] Behind a palisade fence kūmara grow in small mounds of soils blended with pebbles and charcoal, set out neatly in the quincunx pattern noted by early naturalist Joseph Banks. Aute, taro and gourds flourish and several varieties of harakeke (or flax) surround the garden. A tōtara pātaka or kūmara storage house, its earth roof pitched to prevent water from collecting and its timbers stained red

Wiremu Puke demonstrating use of the kō at Te Parapara, Hamilton Gardens, 2012.
Author photo

with ochre produced from rusting hematite, is dried out with a fire at the end of the growing season. Kuia (elderly women) carry out the first planting followed by Wiremu, his whānau and Hamilton gardeners. They use traditional wooden tools – the kō is made from maire and kōwhai – and offer the first harvest of kūmara in the most sacred part of the garden.[136]

Wiremu's efforts at Te Parapara inspired further developments. At Waipatu Marae near Hastings, Arohanui (Hanui) Lawrence (descendant of Hēnare Tomoana, mentioned in Chapter One) began taking her passion for kūmara-growing into her local community in 2006. She asked her cousin if she could put in a garden in a spare paddock at the marae: 'I was wanting to educate the community. But they won't be educated!' she remarked ruefully, adding, 'Now they will, because the sweetcorn is coming on.' Hanui's husband, children and grandchildren share the mahi (work) with her, planting and irrigating by hand. The traditions of her family are being maintained: 'I've grown kūmara all my life. There hasn't been a year when I haven't grown it. We grow our own tipu a traditional way that my parents did ... and rear all our own plants.' Her love for the kūmara is evident. 'When you dig them up it's like digging up pots of gold ... They're so red and rich. No other vegetable affects

Arohanui (Hanui) Lawrence: 'I've grown kūmara all my life.'
Photo: Russell Kleyn

Starting the kūmara tipu at Waipatu Marae.
Photo: Marion Thomson

you like a kūmara does.'[137] Hanui refuses to use chemicals. 'I talk to my grandchildren about weeding the kūmara. [The plants] like to be fondled and touched, like people. I wouldn't like to be sprayed. Would you?' She takes her grandchildren to the gardens to pick the first corn and kamokamo, as her grandmother Kuini did, and uses the garden calendar her grandfather Paraire compiled. Her garden isn't 'manicured': 'I like to leave the crops and the [corn] stalks to dry off, and the birds come. And it's not just any birds – it's yellowhammers, yellow finches and green finches. Great flocks of them, all feeding on the seeds and composting the ground. I just like to leave it.' When her Māori potatoes – Hua Karoro, Moe Moe, Tutaekuri, Karu Parera – came under attack in 2009 and 2010 from psyllid insects, she resisted spraying them. 'I rely on prayer for the potato crop,' she said. 'The psyllid will settle down.' Like wireworm, cabbage white butterflies and aphids, 'they ... have a purpose as well. I don't worry about them too much.' Her approach is about caring for the land. 'It's just something wonderful. That's how I feel about the whenua.'[138] Hanui's marae project was morphing into a Ngāti Kahungunu concept branded as 'Aunty's Garden', an idea centred on her values of kai reka, kai ora and kai wairua (sweet, healthy and spiritual food).

Interest in vegetable varieties grown by early Māori and traditional gardening practices began to surface towards the end of the twentieth century. Tāhuri Whenua, the National Māori Vegetable Growers Collective, was initially set up to support growers' interested in the old potato varieties. The group evolved out of research into early varieties carried out by Nick Roskruge of Massey University. In 2009 a number of kūmara varieties that had vanished from New Zealand but were surviving in Japan were re-introduced and distributed to Māori growers.[139] Hanui was one of these growers. She received them as a 'taonga', blessed and planted them. By January 2010 they were 'thriving'.[140]

The period from the 1960s to the present has seen dramatic changes in the gardening modes of ordinary New Zealanders. In these changes can be seen two principal responses to modernity. The first is reflected in the development of a mass market for standardised garden plants. These plants were grown, and planted, en masse, which meant gardeners started to lose touch with the vast array of plants former generations had been familiar with. At the same time, gardens transformed into play areas for children, rather than spaces in which children worked as part of the household economy. As these processes unfolded, there was a steady decline of horticultural knowledge in the general populace.

The second response to modernity began to become apparent as this generation of children – and *their* children – came of age and acquired homes and families of their own. The de-skilling of the 1960s–80s was met initially by the countercultural movements of the 1970s, but in the 1990s community gardens and community educational programmes started work on re-skilling the population in ways to grow gardens that could meet a range of community needs, from waste reduction and improved nutrition to breaking down social isolation. Old gardening knowledge began to be valued again, and values around developing an 'authentic' connection with the land seemed to be breaking down that old 'antipodean vision' described in Chapter Four. This paved the way for the development of the food sovereignty and food resilience movements that emerged in New Zealand during the watershed years following the Canterbury earthquakes of 2010–12.

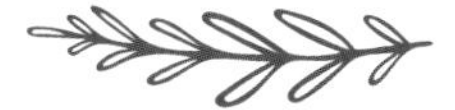

CHAPTER 8

Dreams

In the early twenty-first century a new term has entered public discourse: food sovereignty. It asserts:

> that the people who produce, distribute, and consume food should control the mechanisms and policies of food production and distribution, rather than the corporations and market institutions they believe have come to dominate the global food system.[1]

As the world heads perilously close to a total collapse of food production systems, many gardeners can see how they fit into the bigger picture. The prolonged availability of cheap oil, which keeps our country in motion, is in doubt. As supplies diminish, artificial agricultural fertilisers based on petroleum will not be produced; the trucking industry will be under extreme pressure; and supermarkets and food retailers will struggle to maintain supplies.

But the impending crisis and associated threats to our social system provide fertile ground for optimistic community dreaming and wonderful opportunities for further renewal. As artificial fertilisers become scarce, organic growing methods will be vital. If transporting food across the globe is no longer a viable option, we will need to produce ample food in our own neighbourhoods. As roads empty of traffic, perhaps they may be brought to life with lines of fruit trees, or reconfigured as vegetable corridors. Communities may learn to pool resources, and sharing, once such a normal part of neighbourly life, may be back in vogue. We might find out who lives next door, and install gates in our fences so that we can help each other to grow food, just as residents in Dunedin did in the 1940s.[2]

At the 2010 Ellerslie International Flower Show, exhibitor Carl Pickens, organic landscape designer and former trustee of the Organic Garden City Trust, constructed his garden around the idea of community dreams for a healthy and peaceful planet. His display included 1664 glass bottles, each one containing one person's vision for making the world a better place. Here was a garden of hope and possibilities, a place of dreams, a place where good things would begin. Carl's message was clear: a community approach is vital as we face an uncertain future, somehow gardens hold the key: we must learn to live in spiritual and environmental harmony with nature.

Garden fashion vs garden practice

In this book I have attempted to bury the idea that New Zealand gardens are only worthy of study if they have been designed by a professional, maintained by hired labour, and show off the fashions of the day. When we look closely at the home gardens of ordinary New Zealanders, they turn out to be rich and dynamic. They demonstrate relationships between people and plants, between families and landscapes. Many of these gardeners have, in fact, been rather impervious to fashionable whims. For centuries our gardens have been a place to grow our own food, and this seems only likely to develop further.

A food garden is a means to independence, and the control of others' gardening activities is one manifestation of how power may be wielded in society. Colonial governments sometimes pursued expansionist aspirations through citizen-led horticultural societies that encouraged people first to grow more food, and later to beautify urban environments with ornamental plantings. Garden-making consumed an enormous amount of collective time in early Māori communities; as muskets entered the local economy, food growing developed beyond providing for immediate needs to meeting iwi aspirations for expansion. At the same time, the government's military forces set about destroying those Māori food gardens that constituted a resource that could threaten imperial power. Later still, in the 1950s, when such gardens no longer posed any risk to the government, rural Māori communities were called upon to develop vegetable gardens by health authorities who hoped to curb the incidence of tuberculosis in the general population, in the same way that Māori pā were considered to pose a typhoid risk for Pākehā in the early twentieth century.[3] Meanwhile, Pākehā communities were exposed to greater pressure to keep 'enemies' at bay with a chemical arsenal partly developed in wartime.

It is remarkable that so few New Zealand gardeners caved in to the guilt-inducing inducements put forward by horticultural and beautifying societies and industries keen to sell new products. Scepticism about the genuine virtues of chemical pesticides and herbicides, coupled with an unwillingness to pay for them, meant that many people did not buy into the sterilising vision of mid-twentieth-century 'garden experts'. Many interviewees remembered less toxic or plant-based concoctions in the garden shed, such as superphosphate, Derris Dust and nicotine-based insecticide sprays, but it appears the more potent chlorinated hydrocarbon pesticides (like DDT, dieldrin, adrin, endrin and lindane) and herbicides (like 2,4,5-T and 2,4-D) were less popular.

This is not to say that the good work of horticultural societies and other vehicles of New Zealand's dominant culture has been a waste of time, or that it has been completely ignored. Nor is it to blacken the reputation of the hard-working people in these societies who strove to raise the level of gardening knowledge in our communities. But in the main, New Zealanders tended to adopt new information selectively to suit their existing needs. Māori communities were quick to recognise and harness the value of new crops introduced by Europeans. Pākehā colonists, asked to grow food for the good of the colony, grew plenty; when asked to grow flowers to make their front gardens look pretty, however, they were slower to conform. Only when their food systems were well established did they begin to turn their hand to beautification projects.

The ever-changing gardening methods and styles described in historic newspaper columns and gardening manuals give us the impression that gardening practices changed dramatically every few years, but this is not the case. Interviews with gardeners and examination of gardening diaries, letters and other sources reveal a different story. One gardening innovation proved significant and enduring, however. The compost heap supported the home garden's primary purpose of food production at a time when horse manure was becoming less plentiful, and met particular needs created by World War Two such as the lack of manpower

Roses at Takahanga Marae, Kaikōura, 2010.
Author photo

and artificial fertilisers. Composting also helped legitimise the 'rubbish heaps' of slowly rotting organic material that every gardener had, and which were perceived to attract flies and vermin and thus threatened to drag society into the slums.

The connection between the use of human waste and food gardening is an area ripe for research. The earlier 'abhorrence' reportedly felt by Pākehā at Chinese manuring practices sat uncomfortably alongside the fact that fertilisation of the veggie patch from 'the can' was not uncommon. Māori may have had different practices; manuring fields with human excrement was resisted during the period of early European settlement for reasons of tapu.[4] In the wake of the Canterbury earthquakes of 2010 and 2011, when sewage systems were disrupted, Christchurch residents were actively encouraged to dig human waste into their gardens and make composting toilets.[5]

Resisters

Gardening activity has often been used to mark explicit opposition to a dominant culture, and this can be seen in New Zealand history as much as anywhere. At one end of the spectrum – where most 'resisters' are probably to be found, whether through indolence or ignorance or in protest – is the individual who does not keep a 'good' garden: who lets weeds grow, makes little attempt to cultivate vegetables and flowers and does not maintain a tidy lawn. At the other end are those who provide a clearly articulated political critique of the global economic system and view gardens as places where people may claw back their sovereignty. This is why the early compost movement is littered with communists and social creditors, and attracted hippies and members of the environmentalist Values Party. Māori protesters have used gardening as a way of stating their claim to land, of showing independence, of asserting mana. Gardening as resistance helps us to better understand the notion of gardening as compliance and gives us another tool to explain why people garden.

Kingsland Community Gardens, Auckland, 2010.
Author photo

Few people consciously participate in this level of contestation. They absorb information about how to prune roses and what spray to apply for black spot without questioning the political motivations of the author. The more interesting question, actually, is how many people read such information *and then applied it*.

The whakapapa of gardens

Gardeners frequently know the lineage of their favourite plants: this one came from a cutting from a friend or neighbour, that one has been passed down through the family … The informal sharing of cuttings, tubers, bulbs and seeds is an important way of acquiring plant material and has contributed to the stability of gardening forms through the decades.

Of equal importance is the whakapapa of garden knowledge. The Aotearoa New Zealand garden knowledge economy has largely been drawn from personal experience and passed from generation to generation through families and whānau. Youngsters who were involved in garden chores developed a healthy store of knowledge and skills as a

David Musgrave's remodelled vegetable garden at Waihi Bush in 2009. This 'mandala'-shaped garden was inspired by permaculture design principles. After several generations the form of the garden has changed, but still carries important connections with past eras.

Author photo

result. From the 1960s, however, children's involvement in gardening appears to have tailed off; later in life some struggled with the basics of gardening and gave up in despair, while others simply didn't bother to start. Despite the proliferation of garden centres and gardening publications in the late twentieth century, the culture of maintaining a home garden of any sort was diminishing, a factor that has not been sufficiently explored in garden histories to date.

Although the traditional reliance on garden literature to construct our histories reveals useful detail regarding the gardens of the wealthy, it is generally not helpful in considering more humble gardens. It is important to examine a home garden in the context in which it is situated. Greater recognition of place in our histories would help to illustrate the ways people have created gardens in relation to the social and cultural

Max Podstolski and Clare Reilly in their Cashmere garden, Christchurch, 2016.

Photo: Julie Podstolski

worlds surrounding them, and how garden information has circulated according to social class. It might also help us to recognise the environmental impact our gardening has wrought on New Zealand's ecosystems – not only the physical impacts of our historical use of toxic chemicals, our profligate use of water, or our interference with ecosystems through selected planting schemes (although these impacts and others like them are significant), but also what these interventions mean collectively.

Place-based gardening history also tells us about people's values in respect to the environment. When we learn about the gardens of Māori whānau who have been located in a particular place for generations, we gain appreciation of what is meant by the term kaitiakitanga or guardianship. When we read about how the native plant craze of the early 1930s played out in particular Pākehā gardens – especially those on hilly areas perhaps prone to erosion – we can almost sense those places speaking to their gardeners. When we reflect on the way gardening allows people to develop a relationship with their environment, to become part of it, we gain fresh insight into our history as a nation. Despite the apparent invincibility of the British imperial vision, the land did not always allow for that vision to play out. It could be too wet, too steep, too dry, and such constraints affected and altered the imperial project.

A wonderful description of responses to environmental limits is given by James Beattie in his article 'Science, religion, and drought: Rainmaking experiments and prayers in North Otago, 1889–1911'.[6] Launching explosives into clouds and collective prayers were perhaps desperate attempts to push ahead with a colonial approach to land management that was not working. Gardening is one space where the land moderated, constrained and shaped the vision of British colonists.

A new dream: Edible garden cities

An example of garden dreaming surfaced in Christchurch in the wake of the 2010 and 2011 earthquakes. The quakes destroyed most of the central business district and made large swathes of residential land uninhabitable, affecting around 8000 homes. One suburban stretch along the Avon River – formerly the ancient site of Ōtautahi and William 'Cabbage' Wilson's 1850s garden, the first botanical garden and the city's first plant nursery (see Chapter Two) – had a number of badly damaged houses set in mature gardens. Householders in this suburb were forced to leave their quake-damaged properties, and the future of the area, now labelled the 'red zone', was uncertain.

Concerned citizens and organisations together created a plan for a park that would encompass the abandoned gardens and included new community gardens and allotments, and the group Community Gardening and Food Resilience emerged in 2012 to realise the dream.[7] The group hoped to create an immense inner-city food-producing area: a Transition Town project writ large.

In 2013 I had the opportunity to lead a collaboration between Christchurch organisations that were determined to make more food accessible to more people. The collaboration became the Food Resilience Network (FRN), and included Christchurch City Council, Canterbury District Health Board and numerous non-government organisations. (By 2016 the combined reach of the FRN would be an estimated 11,000 people.[8]) The focus was on sustainable use of the land along the Avon River, and on how to 'retro-fit' food resilience into Christchurch as the city was rebuilt. The group's vision was to create 'a patchwork of food-producing initiatives based around local hotspots and linked together like a ribbon woven into the fabric of our communities'.[9]

In 2014 Christchurch City Council developed its own food resilience policy in which it articulated the vision to be the 'best edible garden city in the world'.[10] In 2015 the FRN successfully submitted an expression of interest to government about developing a community food hub in the central city – a building set in a community orchard and garden to showcase the local movement. It was to be the 'front door' for the food resilience movement, and one node among many that would embed food resilience in communities across the city. The space, named Ōtākaro Orchard, was opened by the prime minister in early 2016, and in 2019 plans for food initiatives throughout the red zone were formally incorporated into the Ōtākaro Avon River Corridor Plan.

Such developments were, once again, informed in part by developments overseas. As noted by anthropologist David Boarder-Giles, participants in the red zone and Ōtākaro Orchard are 'part of a transnational counterpublic of permaculture advocates and practitioners who draw on common texts and discourses', and whose 'founders and organisers are keenly, directly aware of each others' work'.[11] The FRN, and its

Bailey Peryman, an early leader in Christchurch's post-earthquake food resilience scene, turning compost in the central city garden, Agropolis, 2014.

Photo: Eroica Ritchie

flagship project Ōtākaro Orchard, had direct ties with, among others, Beacon Food Forest in Seattle and, to a lesser extent, Incredible Edible Todmorden in the UK.

A dream for the future

At the beginning of the twenty-first century, the dreams of New Zealanders as a gardening people stand awaiting fulfilment. The importance of locally grown food will increase in the coming decades because the global food system is in need of a dramatic and inevitable shake-up.[12]

Gordyn Hamblyn in the transforming rubble of Ōtākaro Orchard, Christchurch, 2018.
Photo courtesy of Peanut Productions

Aunty Aroha Reriti-Crofts blessing Ōtākaro Orchard, Christchurch, 2019.
Photo courtesy of Peanut Productions

Fortunately the number of New Zealanders interested in growing their own is increasing. As Transition Town initiative projects develop, so too will gardening possibilities. Home gardens will brim again with fruit and vegetables, and many of these will be grown from heirloom seeds preserved during the twentieth century by enthusiasts and now rediscovered for their nutritive properties, disease resistance and vibrant colours. Where there is no longer private land available for this work, public space will transform into edible space as people establish food sovereignty.[13] Community gardens and the concept of the 'edible landscape' will become more popular. More community orchards may develop; fruit trees established and maintained by local residents will appear in parks and on spare land; the fringes of sports fields will host raised beds and berry gardens; nut trees will be planted on empty frontages. These changes will be led by private citizens, who will establish food beds along their front fences so that neighbours may access and share produce. It is to be hoped that local councils will see a good thing and follow suit.

As the century unfolds we may well see a decline in the cultural value of the lawn and other grassed areas, partly as a result of intensification: there may be less space for lawns, especially where they compete for priority with food plantings. Monocultural lawns and grass berms, so resource-intensive and with few ecosystem benefits, may be replanted with edible or drought-tolerant plants, or with new kinds of grass that require minimal mowing, irrigation and fertilisation.

It is difficult to imagine any upswing in the kind of gardens earlier promoted by beautifying societies; verdant lawns and technicolour flowerbeds will not be making any dramatic comeback. Instead flowers will likely be intermingled with vegetables and herbs, and celebrated not only for their beauty but also for their contribution to creating vibrant, resilient and healthy ecosystems.

Equally important will be the evolution of local informal gardening exchange networks. Instead of viewing our urban spaces as concrete deserts interspersed with community gardens, we will re-imagine our cities and towns as garden systems, networks of home, park, community, school and marae gardens. Our open spaces will be thought of as foraging networks, mahinga kai; we will know where and when to seek sweet chestnuts, tasty apricots or the lushest sorrel. Perhaps, in time, this new system will morph more formally into a renewed commons.

Pear trees fruiting at Ōtākaro Orchard, Christchurch, 2020.
Photo: Rachel Vogan

This is my dream, and it is the dream of many New Zealanders who hope to increase the capacity of communities to feed themselves by harnessing the potential of gardens, not only in case oil supplies fail, but because it is inherently a good way forward.

The dreams of successive waves of migrants to these shores, many of whom envisaged Aotearoa as a network of abundant food gardens, continue to play out and find new hope in community-based urban food projects. Gardens, ngā māra, have always been at the heart of life in Aotearoa, even if we have not always recognised this. They are as crucial to our wellbeing now as they have ever been.

Glossary of Māori terms

atua – god

aute – paper mulberry

hapū – tribe or subtribe

hoto – shovel

iwi – tribe, people, nation

kāheru – spade

kai – food

kāinga – village

kaumātua – elderly person

ketu – small canoe paddle-shaped tool

kō – digging stick

koru – a spiral form

kuia – elderly woman

mahi – work

mahinga kai – food gathering and growing place

māra – garden

pā – fortified village

pākoro – storage place for potatoes

pātaka – storehouse raised on posts

pūhā – perennial sowthistle

rangatira – chief, boss, owner (male or female)

rohe – boundary, district, region, territory, area, border (of land)

takahi – footrest on kō

tangi – funeral

taonga – treasure, something prized

tapu – sacred, prohibited

Te Ika a Māui – the North Island

te reo Māori – Māori language

Te Wai Pounamu – the South Island

teka – footrest on kō

timo – grubber

tipu – seedling, shoot or bud

tohunga – expert, priest

uwhi – a kind of yam

waka – canoe

whakapakoko – statue, idol, carved image

whakapapa – genealogy, genealogical table, lineage, descent

whānau – extended family, family group

Notes

Introduction

1   Derek Jarman, *Derek Jarman's Garden* (London: Thames and Hudson, 1995), 40.

2   Bernard Smith, *European Vision and the South Pacific* (New Haven: Yale University Press, 1985), 141, 148; Richard Grove, *Green Imperialism: Colonial expansion, tropical island Edens and the origins of environmentalism* (Cambridge: Cambridge University Press, 1995).

3   Miles Fairburn, *The Ideal Society and its Enemies: The foundations of modern New Zealand society 1850–1900* (Auckland: Auckland University Press, 1989), especially 29–42.

4   Austin Mitchell, *The Half-gallon Quarter-acre Pavlova Paradise* (Christchurch: Whitcombe & Tombs, 1972), 111.

5   Ibid., 114.

6   Janet Frame, *Living in the Maniototo* (Auckland: Vintage, 2006 edn), 77, 115.

7   Keri Hulme, *The Bone People* (Auckland: SPIRAL and Hodder and Stoughton, 1985), 76.

8   See, for example, Keith Sinclair, *A History of New Zealand* (rev. edn) (Auckland: Penguin, 2000), 313.

9   James Belich, *Paradise Reforged: A history of the New Zealanders from the 1880s to the year 2000* (Auckland: Allan Lane, 2001), 348.

10 There are a few, however. For example, see references in Barbara Brookes, Annabel Cooper and Robin Law (eds), *Sites of Gender: Women, men & modernity in Southern Dunedin, 1890–1939* (Auckland: Auckland University Press, 2003); Matt Morris, 'A History of Christchurch Home Gardening from Colonisation to the Queen's Visit: Gardening culture in a particular society and environment' (PhD thesis, University of Canterbury, 2006); T. Strongman, *The Gardens of Canterbury* (Wellington: A.H. & A.W. Reed, 1984).

11 Michael King, *The Penguin History of New Zealand* (Auckland: Penguin, 2003), 428.

12 For initial comments in this direction, see K. Raine and J. Adam, 'The Settlers' Gardens', in Matthew Bradbury (ed.), *A History of the Garden in New Zealand* (Auckland: Viking, 1995), 70; and K. Raine, 'Domesticating the Land: Colonial women's gardening', in Bronwyn Dalley and Bronwyn Labrum (eds), *Fragments: New Zealand cultural and social history* (Auckland: Auckland University Press, 2000).

13 See, for example, John Dixon Hunt and Joachim Wolschke-Bulmahn (eds), *The Vernacular Garden* (Washington, DC: Dumbarton Oaks, 1993); Michel Conan, 'From Vernacular Gardening to a Social Anthropology of Gardening', in Conan (ed.), *Perspectives on Garden Histories* (Washington, DC: Dumbarton Oaks, 1999); Katie Holmes, '"I have built up a little garden": The vernacular garden, national identity and a sense of place', *Studies in the History of Gardens and Designed Landscapes*, vol. 21, no. 2, 2001, 115–21; John Hammetter, 'Gardening and Meaningful Lives: An ethnography of Milwaukee-area residential vernacular gardens' (PhD thesis, Northwestern University, 2002).

14   Margaret Willes, *The Gardens of the British Working Class* (New Haven: Yale University Press, 2014), 2.

15   See Jim McAloon, *No Idle Rich: The wealthy in Canterbury & Otago, 1840–1914* (Dunedin: University of Otago Press, 2002), 25–26, 31–32, for discussion on the complexities of class and power and difficulties with categorisation in the colonial New Zealand context.

16   In doing so, I hope this work builds on earlier works of New Zealand garden history, notably Helen Leach, *1,000 Years of Gardening in New Zealand* (Wellington: Reed, 1984); Mathew Bradbury (ed.), *A History of the Garden in New Zealand* (Auckland: Penguin, 1995); and Bee Dawson, *A History of Gardening in New Zealand* (Auckland: Random House, 2012).

1 New food in the gardens: 1800–50

1   Rāwiri Taonui, 'Canoe traditions', Te Ara – the Encyclopedia of New Zealand: www.TeAra.govt.nz/en/canoe-traditions

2   Dallas Matoe, interviewed 10 April 2010.

3   'Horticulture', Exhibit at Puke Ariki, New Plymouth, March 2010.

4   Matthew Prebble et al., 'Early Tropical Crop Production in Marginal Subtropical and Temperate Polynesia', *Proceedings of the National Academy of Sciences of the United States of America*, vol. 116, no. 18, 30 April 2019, 8824–33.

5   Louise Furey, 'Maori Gardening: An archaeological perspective' (Wellington: Science and Technical Publishing, Department of Conservation, 2006).

6   Ibid., 6.

7   Matthew Prebble et al., 'Early Tropical Crop Production in Marginal Subtropical and Temperate Polynesia', *Proceedings of the National Academy of Sciences of the United States of America*, vol. 116, no. 18, 30 April 2019, 8824–33.

8   Ibid., 8832.

9   Helen Leach, *1000 Years of Gardening in New Zealand* (Wellington: Reed, 1984), 10, 43.

10   Furey, 'Maori Gardening', 60.

11   Janet Davidson, 'Auckland prehistory: A review', Auckland Institute and Museum Records, 1 December 1978, 4; 11.

12   Ibid., 6.

13   www.doc.govt.nz/conservation/historic/by-region/auckland/central-and-south-auckland/otuataua-stonefields/; www.manukau.govt.nz/EN/Yourcommunity/ParksWalksBeaches/FindAPark/Pages/OtuatauaStonefields.aspx

14   Paul Moon, *The Struggle for Tāmaki Makaurau: The Māori occupation of Auckland to 1820* (Auckland: David Ling, 2007), 46–47. See also Susan Bulmer, 'Sources for the archaeology of the Maaori settlement of the Taamaki volcanic district' (Wellington: Department of Conservation, 1994), 41–42.

15   Moon, *The Struggle for Tāmaki Makaurau*, 51; Bulmer, 'Sources', 16.

16   Bulmer, 'Sources', 27.

17   Furey, 'Maori Gardening', 23.

18   Leach, *1000 Years of Gardening in New Zealand*, 43.

19 Ibid.

20 Susan Bulmer, 'Ngā māra: Traditional Māori gardens', in Matthew Bradbury (ed.), *A History of the Garden in New Zealand* (Auckland: Penguin, 1995), 28–30.

21 Ibid.

22 Auckland War Memorial Museum exhibit.

23 Leach, 1000 *Years of Gardening in New Zealand*, 50.

24 Thanks to Frances Helps for this insight, 10 December 2011.

25 Hare Te Heihei, quoted in Jeffrey Sissons and Wiremu Wi Hongi, *Ngā Pūriri o Taiamai* (Auckland: Reed, 2001), 21.

26 Prebble et al., 'Early Tropical Crop Production', 8828.

27 Ibid., 8825.

28 W.A. Taylor Papers, Box 1, Folder 7, 47–48, CM; see also Te Maire Tau, 'Ngai Tahu Claim, WAI-27: Mahinga Kai (Tuahiwi)' (Tuahiwi Marae, 1988), 14.

29 Terry Ryan, 'Ti Kouka Whenua': http://library.christchurch.org.nz/TiKoukaWhenua/

30 Harry Evison, *Te Wai Pounamu: The greenstone island* (Wellington: Aoraki Press, 1993), 5; Tau, 'Ngai Tahu Claim', 13.

31 Philip Simpson, *Dancing Leaves: The Story of New Zealand's Cabbage Tree* (Christchurch: Canterbury University Press, 2000), 145, 150.

32 Bulmer, 'Ngā māra', 21.

33 Elsdon Best, *Maori Agriculture* (Wellington: A.R. Shearer, 1976), 278.

34 Wiremu Puke, interviewed 19 January 2012.

35 Best, *Maori Agriculture*, 278; Bulmer, 'Ngā māra', 22.

36 John Clemens, '*Clianthus*: Veiled in mystery', *New Zealand Garden Journal* (Journal of the Royal New Zealand Institute of Horticulture), vol. 5, no. 1, June 2002, 4–5: www.rnzih.org.nz/pages/Clianthus.htm

37 McGregors Seed Catalogue: www.mcgregors.co.nz/products/showprod. php?sec=3&cat=NZNA&prod=M4400

38 Clemens, '*Clianthus*', 4–5.

39 Ibid.

40 Auckland War Memorial Museum exhibit.

41 Wiremu Puke, interviewed 19 January 2012.

42 Best, *Maori Agriculture*, 28, 34.

43 Auckland War Memorial Museum exhibit.

44 Best, *Maori Agriculture*, 100.

45 Richard Cruise, *Journal of a Ten Months' Residence in New Zealand* (1820) (Christchurch: Pegasus Press, 1957 edn), 84.

46 Ernst Dieffenbach, *Travels in New Zealand* (London: John Murray, 1843), 48, 49.

47 Best, *Maori Agriculture*, 117.

48 Ibid., 118–19.

49 Auckland War Memorial Museum exhibit.

50 Ibid.

51 'Horticulture', Exhibit at Puke Ariki, New Plymouth, March 2010.

52 Wiremu Puke, interviewed 19 January 2012.

53 Samuel Marsden in John Rawson Elder (ed.), *The Letters and Journals of Samuel Marsden 1765–1838* (Dunedin: Coulls Somerville Wilkie and A.H. Reed, 1932), 179.

54 Elsdon Best, *The Astronomical Knowledge of the Māori: Genuine and empirical* (Christchurch: Kiwi Publishers, 2002), 29. The links that Māori made between what was happening astronomically and their garden-making activities have led some to suggest a comparison with the later work of biodynamic gardeners.

55 Canon Stack, quoted in Best, *Maori Agriculture*, 23.

56 William J. Phillips, *Māori Life and Custom* (Auckland: A.H. & A.W. Reed, 1966), referred to by Martin Tickner in *Southern Seed Exchange Newsletter*, no. 31, September 2010.

57 Atholl Anderson, *The Welcome of Strangers: An ethnohistory of Southern Maori AD1650–1850* (Dunedin: University of Otago Press, 1998), 128.

58 J. Cook, quoted in Best, *Maori Agriculture*, 22.

59 Joseph Banks, quoted in ibid., 28–29.

60 Crozet, quoted in ibid., 279.

61 Forster, quoted in ibid., 280; https://teara.govt.nz/en/kai-pakeha-introduced-foods/page-1

62 Edward J. Wakefield, *Adventure in New Zealand from 1839 to 1844*, vol. 1 (London: John Murray, 1845), 24.

63 H. Hanson Turton to the Wesleyan Missionary Secretaries, 5 January 1847, Wesleyan Missionary Society, Folder 43, Methodist Church of New Zealand Archives.

64 Cook, quoted in Best, *Maori Agriculture*, 281.

65 Savage, quoted in ibid., 282.

66 Note in Elder (ed.), *The Letters and Journals of Samuel Marsden 1765–1838*, 527.

67 Marsden, in ibid., 100.

68 Ibid., 95, 230.

69 Atholl Anderson, 'Mahinga kai: Evidence for the Ngai Tahu Claim to the Waitangi Tribunal' (Tuahiwi, 1988), 8, 9.

70 Ibid., 12.

71 Ibid., 19.

72 Ibid., 21.

73 Ibid., 33, 34; Wakefield, *Adventure in New Zealand from 1839 to 1844*, 25.

74 Wiremu Puke, interviewed 19 January 2012.

75 R. Stone, *From Tamaki-Makau-Rau to Auckland* (Auckland: Auckland University Press, 2001), 70.

76 Jeffrey Sissons and Wiremu Wi Hongi, *Ngā Pūriri o Taiamai* (Auckland: Reed, 2001), 24. John Nicholas was Samuel Marsden's friend and travelling companion, and wrote the account of Marsden's visit to New Zealand 1814–15.

77 Cruise, *Journal of a Ten Months' Residence in New Zealand*, 148.

78 Ibid., 185, 195, 217–18.

79 Augustus Earle, *A Narrative of a Nine Months' Residence in New Zealand in 1827* (Christchurch: Whitcombe & Tombs, 1909), 75; Dumont d'Urville, *New Zealand 1826–1827*, trans. Olive Wright (Wellington: Wingate, 1950), 186; Te Urewera, WAI 894 (Wellington: Waitangi Tribunal, 2009), 157; Anita Miles, 'Te Urewera', quoted in Te Urewera, 228; Te Urewera, 204.

80 Dieffenbach, *Travels in New Zealand*, 142, 170.

81 Felton Mathew, 4 February 1840, in J. Rutherford (ed.), *The Founding of New Zealand* (Wellington: A.H. and A.W. Reed, 1940), 31.

82 Mathew, 16 February 1840, in Rutherford (ed.), *The Founding of New Zealand*, 47–48.

83 Mathew, 27 February 1840, in ibid., 65.

84 John B. Williams, Journal: 1842–1844, NZMS 1063, Auckland Libraries, Sir George Grey Special Collections.

85 Stone, *From Tamaki-Makau-Rau to Auckland*, 69.

86 Earle, *A Narrative of a Nine Months' Residence in New Zealand*, 76.

87 Marsden, in Elder (ed.), *The Letters and Journals of Samuel Marsden 1765–1838*, 125.

88 Ibid., 99.

89 John Nicholas, quoted in Sissons and Wi Hongi, *Ngā Pūriri o Taiamai*, 22.

90 Frederick Spencer, 'Reminiscences of an old New Zealander', NZMS 865/1, Auckland Libraries, Sir George Grey Special Collections, 151.

91 Marsden, in Elder (ed.), *The Letters and Journals of Samuel Marsden 1765–1838*, 97.

92 Cruise, *Journal of a Ten Months' Residence in New Zealand*, 52.

93 Ibid., 168.

94 Anderson, 'Mahinga Kai', 78.

95 Ibid.

96 H. Hanson Turton to the Wesleyan Missionary Secretaries, 26 May 1841, Wesleyan Missionary Society, Folder 43, Methodist Church Archives.

97 Harry Evison, *The Long Dispute: Māori land rights and European colonisation in southern New Zealand* (Christchurch: Canterbury University Press, 1997), 177; see also Anderson, *The Welcome of Strangers*, 151–52.

98 John Robert Godley to his father, 9 April 1850, J.R. Godley and E.G. Wakefield Correspondence, box 3, vol. 3, 466, 475/50, CM.

99 Shirley Tunnicliff (ed.), *The Selected Letters of Mary Hobhouse* (Wellington: Daphne Brasell Associates Press, 1992), 40–41.

100 Sarah Mathew, 'Autobiography, 1873', in J. Rutherford (ed.), *The Founding of New Zealand* (Wellington: A.H. and A.W. Reed, 1940), 201; William Swainson, quoted in Stone, *From Tamaki-Makau-Rau to Auckland*, 290.

101 Anderson, 'Mahinga Kai', 23.

102 Edward Ward, 'The Ward diary', *Press*, 14 February 1925.

103 Marsden, in Elder (ed.), *The Letters and Journals of Samuel Marsden 1765–1838*, 202.

104 Ibid., 19 August 1844; 2, 4, 6 September 1844.

105 George Metehau, 'To the white people', *LT*, 12 June 1852.

106 Charles Heaphy, *Narrative of a Residence in Various Parts of New Zealand* (London: Smith, Elder, & Co., 1842), 30–31.

107 John Cracroft Wilson, Diary/Reminiscences, c. 1854, ARC 1991.53, Canterbury Museum, 78.

108 Wellington Tenths Trust, *GIS Map Book, 2004* (Wellington: Borcoda Print, 2004), 12.

109 Terry Ryan, interviewed 25 September 2004.

110 'Hēnare Tomoana', Ministry of Culture and Heritage: https://nzhistory.govt.nz/ people/henare-tomoana

111 Arohanui Lawrence, interviewed 21 January 2010.

112 'Fishing by the moon', Arohanui Lawrence, private collection.

113 Robert Cooper, 'The Banks Lecture: Early Auckland gardens', *Journal of the Royal New Zealand Institute of Horticulture*, 1971, 100.

114 For example, Bee Dawson, *A History of Gardening in New Zealand* (Auckland: Random House, 2010).

2 The fruitful paradise: 1840–1914

1 *New Zealand Colonist and Port Nicholson Advertiser*, 23 September 1842, 4.

2 S. Challenger, 'Pioneer Nurserymen of Canterbury, New Zealand (1859–65)', *Garden History*, vol. 7, no. 1, 1979, 27.

3 William Tye, Diary, 1838–1848: 29 October 1842, 9 November 1842, 29 July 1844, 10 August 1844, 24 August 1844, 10 September 1844, 16 September 1844, MS-2190, ATL.

4 Charles Heaphy, *Narrative of a Residence in Various Parts of New Zealand* (London: Smith, Elder & Co., 1842), 31–32.

5 Ibid., 31.

6 Ibid.

7 Ibid., 31–32.

8 John Williams, Journal: 1842–1844, NZMS 1063, Auckland Libraries, Sir George Grey Special Collections.

9 Ibid.

10 Vaughan Wood, 'Appraising soil fertility in early colonial New Zealand: The 'biometric fallacy' and beyond', *Environment and History*, vol. 9, no. 4, 2003, 397; Frances Loeffler, 'Pteridomania: A visual history of the fern in New Zealand' (MA thesis, Victoria University of Wellington, 2006), 38–39.

11 Henry Hanson Turton to the Wesleyan Missionary Secretaries, 12 November 1842, Wesleyan Missionary Society, Folder 43, Methodist Church Archives.

12 Te Maire's rationale for this translation follows: pu – a clump of forest, tari – a noose, and motu – to cut the noose. In this reading, 'nga' is simply a passive. Other readings have assumed that the middle section should be read as 'taringa', or ear. Te Maire Tau, email to author, 10 February 2006. Graham M. Miller, 'Deans, John and

Deans, William', Dictionary of New Zealand Biography, Te Ara – the Encyclopedia of New Zealand: www.teara.govt.nz/en/biographies/1d7/deans-john

13 Robert Harris, 'As It Was: Early Māori and European settlement', in S. Owen (ed.), *The Estuary: Where our rivers meet the sea* (Christchurch: Parks Unit, Christchurch City Council, 1992), 14.

14 Te Maire Tau, *Cultural Report on the Southwest Area Plan for the Christchurch City Council* (Christchurch: Christchurch City Council, 2005), 20.

15 John Deans to John Deans Senior, 10 January 1844, in John Deans (ed.) *Pioneers of Canterbury: Deans letters 1840–1854* (Christchurch: Cadsonbury, 1997), 81.

16 Ibid., 28 September 1845, 91.

17 Ibid., 20 December 1847, 118.

18 Ibid., 8 December 1849, 157.

19 Ibid., 17 February 1851, 202; 8 February 1853, 239.

20 Ibid., 21 March 1853, 245–46; John Deans to James Deans, 27 October 1853, 263.

21 Ibid., 2 January 1854, 271.

22 John Robert Godley to his father, 9 April 1850, in J.R. Godley and E.G. Wakefield Correspondence, box 3, vol. 3, 455, 475/50, CM.

23 Charlotte Godley, 4 March 1851, in Charlotte Godley, *Letters from Early New Zealand by Charlotte Godley 1850–1853* (Christchurch: Whitcombe & Tombs, 1951), 182.

24 Ibid.

25 Ibid., Charlotte Godley to her mother, 14 January 1852, 287.

26 John Cracroft Wilson, Diary/Reminiscences, c. 1854, 84, ARC 1991.53, Canterbury Museum.

27 'A General Summary of the Census of Population from the Kaikouras to the Waitaki, not including the Akaroa & Pigeon Bay Districts', in James Hight and C. Straubel (eds), *A History of Canterbury*, vol. 1, Appendix V (Christchurch: Whitcombe & Tombs, 1957).

28 G.R.H., 'What Christchurch was like in 1852', *Press*, 20 January 1900.

29 H. Evison, *Te Wai Pounamu: The Greenstone Island: A history of the Southern Maori during the European colonisation of New Zealand* (Wellington: Aoraki Press, 1993), 334.

30 *LT*, 14 June 1851.

31 *LT*, 20 September 1851.

32 *LT*, 18 October 1851.

33 William Trotter, *Gardening Diary*, September 1851, MS-2163, ATL.

34 Trotter, *Gardening Diary*, January 1855, MS-2163, ATL.

35 Trotter, *Gardening Diary*, 26 April 1855, MS-2163, ATL.

36 James Beattie and John Stenhouse, 'Empire, environment and religion: God and the natural world in nineteenth-century New Zealand', *Environment & Society*, 13 (2007), 413–46.

37 For more on Trotter, see Katherine Raine, 'The Settlers' Gardens', in Matthew Bradbury (ed.), *A History of Gardening in New Zealand* (Auckland: Penguin, 1995), 75–77.

38 www.teara.govt.nz/en/ngai-tahu/page-8

39 Edward Jollie, *Recollections 1841–1863*, 68, ARC 1900.285, Folder 1382, CM.

40 W.G. Brittan, 'The harvest', *LT*, 19 March 1853.

41 On this, see Johannes C. Andersen, *Old Christchurch in Picture and Story* (Christchurch: Simpson and Williams, 1949), 219; Sydney Challenger, '1852? or 1851? Puzzles concerning the Horticultural Society, the Botanical Garden and William Wilson', *CB*, April/May 1978, 18–20; Sydney Challenger, 'William Wilson: Landowner and early nurseryman', *Press*, 20 May 1978; Sydney Challenger, 'Studies on Pioneer Canterbury Nurserymen (1): William Wilson, *Annual Journal of the Royal New Zealand Institute*, no. 6, 1978, 139–62. See also Eric Pawson, 'Confronting Nature', in Graeme Dunstall and John Cookson (eds), *Southern Capital: Christchurch: Towards a city biography* (Christchurch: Canterbury University Press, 2000), 67.

42 Matt Morris, 'A History of Christchurch Home Gardening from Colonisation to the Queen's Visit: Gardening culture in a particular society and environment' (PhD thesis, University of Canterbury, 2006).

43 *LT*, 13 September 1851.

44 *LT*, 4 October 1851.

45 *LT*, 13 December 1851.

46 William Wilson, *LT*, 14 August 1852.

47 Māori Land Court Ikaroa District South Island Minute Book, vol. 1B, 54, CCL.

48 Te Maire Tau, Anake Goodall, David Palmer and Rakiihia Tau, *Te Whakatau Kaupapa: Ngai Tahu Resource Management Strategy for the Canterbury Region* (Wellington: Aoraki Press, 1992); Alexander Mackay, *A Compendium of Official Documents Relative to Native Affairs in the South Island*, vol. 2 (Nelson: Government of New Zealand, 1872), 208.

49 See also Matt Morris, 'Cabbage Wilson's Garden: Overwriting indigenous space in Christchurch, New Zealand, 1850–1856', in Alex Gerbaz and Robyn Mayes (eds), *Palimpsests: Transforming communities* (Perth: Division of Humanities, Curtin University, 2005), 95–107.

50 William Wilson, 'Calendar of garden and farm operations for June', *CGCA*, 3 June 1852.

51 Ibid.

52 Wilson, 'Calendar of garden and farm operations for August', *CGCA*, 5 August 1852; 'Calendar of garden and farm operations for July', *CGCA*, 8 July 1852.

53 Wilson, 'Calendar of garden and farm operations for August', *CGCA*, 5 August 1852.

54 On the topic of medicinal plants in domestic gardens in New Zealand, see Joanna Bishop, 'The Role of Medicinal Plants in New Zealand's Settler Medical Culture, 1850s–1920s' (PhD thesis, University of Waikato, 2014).

55 Wilson, 'Calendar of garden and farm operations for September', *CGCA*, 2 September 1852. 'Convolvulus Major' was probably a reference to *Ipomoea purpurea* (morning glory) and not *Convolvulus arvensis*, later to become the bane of many New Zealand gardeners.

56 Charles Bridge, Diary 1850–1865, 2 July 1852, 29 July 1852, 8 August 1852, 2 September 1852, 23 September 1852, ARCH 205, CCL.

57 'The suburbs of Christchurch', *LT*, 17 April 1852.

58 Ibid.

59 Ibid.

60 Conway Rose, 'Account of the Canterbury Settlement of New Zealand', 7, Canterbury Museum.

61 Robert B. Paul, *Letters from Canterbury* (London: Rivington, 1857), 58–59.

62 John Cracroft Wilson, Diary/Reminiscences, c. 1854, 28, ARC 1991.53, Canterbury Museum.

63 Edward Gibbon Wakefield to Catherine Wakefield, 24 March 1853, J.R. Godley and E.G. Wakefield Correspondence, box 1, vol. 1, 248, 475/50, CM. Edward Gibbon Wakefield to Henrietta Rintoul, 25 March 1853, J.R. Godley and E.G. Wakefield Correspondence, box 1, vol. 1, 251, 475/50, CM.

64 Edward Gibbon Wakefield to Robert Rintoul, 16 April 1853, J.R. Godley and E.G. Wakefield Correspondence, box 1, vol. 1, 254, 475/50, CM.

65 Robert Cooper, 'The Banks Lecture: Early Auckland gardens', *Journal of the Royal New Zealand Institute of Horticulture*, 1971, 100.

66 'Botanical Society', *CGCA*, 1 July 1852.

67 Helen Leach, *1000 Years of Gardening in New Zealand* (Wellington: Reed, 1984), 115; 'Botanical Society', *CGCA*, 1 July 1852.

68 Edward J. Wakefield, *Adventure in New Zealand from 1839 to 1844*, vol. 2 (London: John Murray, 1845), 320.

69 'Horticultural exhibition', *LT*, 7 August 1852; 'Canterbury Horticultural and Agricultural Society', *LT*, 16 April 1853.

70 Auckland Horticultural Society, 'Rules and Schedule of the Auckland Horticultural Society, 1861' (Auckland, 1861), 7–9, GNZ 635, Auckland Libraries, Sir George Grey Special Collections.

71 Auckland Horticultural Society, 'Rules and Schedule of the Auckland Horticultural Society, 1862' (Auckland, 1862), 9–12, GNZ 635, Auckland Libraries, Sir George Grey Special Collections.

72 William Wilson, 'Calendar of garden and farm operations for August', *CGCA*, 5 August 1852.

73 James Beattie, 'The Empire of the Rhododendron: Reorienting New Zealand garden history', in Eric Pawson and Tom Brooking (eds), *Making a New Land: Environmental histories of New Zealand* (Dunedin: Otago University Press, 2013), 241–57.

74 Nigel Rigby, 'The Politics and Pragmatics of Seaborne Plant Transportation, 1769–1805', in Margarette Lincoln (ed.), *Science and Exploration in the Pacific: European voyages to the southern oceans in the eighteenth century* (Woodbridge, Suffolk: Boydell Press, 1998), 97.

75 'The horticultural exhibition', *Press*, 10 March 1865.

76 'The horticultural exhibition', *Press*, 11 March 1865.

77 Auckland Horticultural Society, 'Report and Financial Statement' (Auckland, 1872), 7–9, GNZ 635, Auckland Libraries, Sir George Grey Special Collections.

78 John Adam, 'The leisured classes of Ponsonby: Gardens and parks', in *The New Zealand Map Society Journal*, no. 13, 2000, 47–56.

79 For the full analysis, see Morris, 'A History of Christchurch Home Gardening from Colonisation to the Queen's Visit'.

80 Some of the common vegetables included cabbages, cauliflowers, carrots, potatoes and Victoria rhubarb. Alexander Stuart, 23 August 1864, 'Stuart, Alexander: Diary and Recipe Book', Misc-MS-0607, Hocken Collections, Te Uare Taoka o Hākena, University of Otago.

81 'Catalogue of vegetable, flower, and agricultural seeds, sold by Thomas White' (Wellington, 1866), GNZ 635 W 61, Auckland Libraries, Sir George Grey Special Collections.

82 J. Stanley Monck, Diary, July 1869, ARCH 486, CCL.

83 Monck, 17 August 1869, 25 August 1869, 30 October 1869, 28 November 1869, 4 December 1869, ARCH 486, CCL.

84 Monck, 2, 3 April 1871; 3 October 1871, ARCH 486, CCL.

85 Monck, 20 November 1871, ARCH 486, CCL.

86 Monck, 29 April 1872, 2 July 1873, 5 September 1873, 13–16 and 19 October 1874, 10 June 1875, ARCH 486, CCL.

87 Monck, 19, 21, 26, 28, 29 June 1875, ARCH 486, CCL.

88 Station Gardener's Diary, September 1873, MSY-4832, ATL.

89 Station Gardener's Diary, undated (July? 1876), MSY-4833, ATL.

90 Station Gardener's Diary, 31 November 1876, MSY-4832, ATL.

91 Scrapbook, MS-Papers-8809, ATL.

92 These included large yellow peach, Burbank peach, fig, Prince Egglebert (a pear?), greengage, Golden Esperance plum, nectarine plum, Louise Bonne of Jersey pear, Beurre Diel pear, W.B. Cretien pear, quince, George IV pear, Prince of Wales pear, Irish Peach apple, Bismark, Bismark Stone, Pipin Stone and Gravenstein apple. Scrapbook, MS-Papers-8809, ATL.

93 *Otago Witness*, 14 May 1896, 27.

94 *Press*, 14 January 1905, 21 January 1905.

95 'One hundred years ago: Onehunga Horticultural Society autumn show', *New Zealand Herald*, 20 April 1990, 1:9.

96 Annette Bainbridge, 'Birth, Death, and Marriage in the Garden: Canterbury colonial women gardeners, 1850–1914' (MA thesis, University of Waikato, 2015), 195.

97 Ibid., 95.

98 'Chinese market gardening', *Otago Witness*, 1 June 1878.

99 Joanna Boileau, *Chinese Market Gardening in Australia and New Zealand: Gardens of prosperity* (London: Palgrave Macmillan, 2017), 78.

100 Lily Lee and Ruth Lam, '陈达枝 Chan Dah Chee (1851–1930): Pioneer Chinese market gardener and Auckland businessman', *Environment and Nature in New Zealand*, vol. 6, no. 1, July 2011, 34.

101 James Beattie and Joanna Boileau, '"Cultivated with great carefulness": Chinese market gardening, urban food supplies and public health in Australasia, 1860s–1950s', forthcoming, 2020.

102 Robert Guyton, interviewed 19 March 2020. Robert received this information from a woman who was a child around the end of the Chinese mining at Round Hill, at the start of the twentieth century.

103 'The sewage', *Evening Standard*, 18 April 1878.

104 'Chinese gardens and typhoid', *Auckland Star*, 23 April 1887. See also 'John and his filthy vegetables', *Observer*, 18 May 1895 (reporting on Auckland); 'Chinese market gardens', *Lyttelton Times*, 19 November 1897 (reporting on Wellington); 'Chinese gardens', *Evening Star*, 18 June 1898 (reporting on Dunedin).

105 'Gospel of gardening', *Ashburton Guardian*, 23 August 1910.

106 Boileau, *Chinese Market Gardening in Australia and New* Zealand, 41.

107 'Chinese gardeners retaliating', *Thames Star*, 30 January 1888; 'City Police Court: Larceny', *Otago Daily Times*, 15 May 1890; 'Pahiatua notes', *Hawke's Bay Herald*, 4 July 1899; 'Robbing Chinese gardens', *Inangahua Times*, 12 February 1909; 'Supreme Court: Today's business: Alleged theft of vegetables', *Wanganui Herald*, 30 May 1917; 'Hastings Magistrates Court: A mean theft', *Hastings Standard*, 25 September 1918; 'Magistrate's Court: Alleged theft of cauliflowers', *Taranaki Herald*, 17 October 1918; 'Theft from Chinese garden', *Manawatu Standard*, 23 August 1919; *Evening Post*, 6 July 1921; 'Police Court news: Escapade at Tamaki', *New Zealand Herald*, 8 August 1925; 'Chinese garden raid', *New Zealand Herald*, 23 September 1925; 'Robbed a Chinaman: Thames resident convicted', *Manawatu Standard*, 23 February 1926; 'Theft from Chinese: Youth sent to borstal institute', *Press*, 9 September 1926; 'Vegetable garden robbed', *Otago Daily Times*, 17 December 1928; 'Chinese garden raided', *Manawatu Standard*, 27 December 1929; 'Theft from Chinese', *New Zealand Herald*, 26 August 1937.

108 'Theft from gardens: Offence at Panmure', *New Zealand Herald*, 19 November 1935.

109 'Police Court: Assaulting a Chinaman', *Auckland Star*, 29 May 1882; 'Police Court: Assault', *Thames Advertiser*, 19 March 1888; 'Magistrate's Court', *Nelson Evening Mail*, 30 September 1895; 'Law and police: Assault on a Chinaman', *New Zealand Herald*, 1 March 1898; 'Sunday rioters: Young men heavily fined', *Auckland Star*, 2 September 1903.

110 James Belich, *The New Zealand Wars and the Victorian Interpretation of Racial Conflict* (Auckland: Auckland University Press, 1998 edn), 75.

111 Harry Evison, *The Long Dispute: Māori land rights and European colonisation in southern New Zealand* (Christchurch: Canterbury University Press, 1997), 271.

112 Ibid., 271.

113 Wiremu Puke, interviewed 19 January 2012.

114 Dallas Matoe, interviewed 10 April 2010.

115 'Doctor Donald (Resident Magistrate Lyttelton) to Provincial Secretary – Europeans thieving from native gardens at Rapaki – 7/12/1864', CH287, ICPS 2692/1864; R22196138, Archives New Zealand/Te Rua Mahara o te Kāwanatanga, Christchurch Regional Office.

116 Frederick Spencer, 'Reminiscences of an Old New Zealander', NZMS 865/1, Auckland Libraries, Sir George Grey Special Collections, 57–58.

117 Ibid., 38, 57.

118 *Te Urewera*, WAI 894 (Wellington Waitangi Tribunal, 2009), 228; James Belich, 'Whitmore, George Stoddart', Dictionary of New Zealand Biography, Te Ara – the Encyclopedia of New Zealand: www.teara.govt.nz/en/biographies/1w21/whitmore-george-stoddart

119 Colonel Whitmore, quoted in *Te Urewera*, 327.

120 Donald McLean, quoted in *Te Urewera*, 365.

121 *Te Urewera*, 370.

122 Ibid., 382.

123 Hamlin, 8 June 1870, quoted in ibid., 381.

124 Ibid., 382.

125 Ibid., 409.

3 The City Beautiful and Garden City movements: 1900–40

1 James Beattie, 'Thomas McDonnell's Opium: Circulating, plants, patronage, and power in Britain, China and New Zealand, 1830s–1850s', in Sarah Burke Cahalan and Yota Basaki (eds), *The Botany of Empire in the Long Eighteenth Century* (Washington, DC: Dumbarton Oaks/Harvard University Press, 2017), 180.

2 Sarah Mathew, 3 October 1840, in J. Rutherford (ed.), *The Founding of New Zealand* (Wellington: A.H. and A.W. Reed, 1940), 193, 205–06.

3 Alice Porter to Will Porter, 14 July 1843, Correspondence, NZMS 1256, Auckland Libraries, Sir George Grey Special Collections.

4 Shirley Tunnicliff (ed.), *The Selected Letters of Mary Hobhouse* (Wellington: Daphne Brasell Associates Press, 1992), 57.

5 Ibid., 69.

6 Ibid., 71.

7 Auckland Horticultural Society Report and Financial Statement, 1873 (Auckland, 1873), GNZ 635, Auckland Libraries, Sir George Grey Special Collections.

8 Timaru Beautifying Association, Minute Book 1899–1910, MS-2149, ATL; Wadestown Cottage Garden and Beautifying Society, Minute Book 1918–1922, MSX-3512, ATL.

9 Canterbury United Horticultural Society Minutes, 13 July 1903, CHS Minute Books, unnumbered volume (1903–1908), CHS.

10 First Annual Report of the Canterbury United Horticultural Society, 13 June 1904, CHS Minute Books, unnumbered volume (1903–1908); 'Canterbury Horticultural Society', unlabelled newspaper cutting pinned to minutes for 1907, CHS Minute Books, unnumbered volume (1903–1908), CHS.

11 Annabel Cooper and Marian Horan, 'Down and Out on the Flat: The gendering of poverty', in B. Brookes, A. Cooper and R. Law (eds), *Sites of Gender: Women, men & modernity in Southern Dunedin, 1890–1939* (Auckland: Auckland University Press, 2003), 190.

12 'Editorial', *Press*, 30 April 1900.

13 James Beattie, *Empire and Environmental Anxiety: Health, science, art and conservation in South Asia and Australasia, 1800–1920* (New York: Palgrave Macmillan, 2011), 65.

14 Ibid., 92–94.

15 Barbara Brookes, 'The Risk to Life and Limb: Gender and health', in Brookes, Cooper and Law (eds), *Sites of Gender*, 289; W. David McIntyre, 'Outwards and

Upwards: Building the city', in Graeme Dunstall and John Cookson (eds), *Southern Capital: Towards a city biography, 1850–2000* (Christchurch: Canterbury University Press, 2000), 102.

16 David McGill, *Lower Hutt: First Garden City* (Lower Hutt: Lower Hutt City Council, 1991).

17 Edward J. Wakefield, *Adventure in New Zealand from 1839 to 1844*, vol. 1 (London: John Murray, 1845), 146. The Taranaki Exhibit at the 1906 International Exhibition in Christchurch also promoted Taranaki as 'The Garden of New Zealand'. See https://teara.govt.nz/en/zoomify/25366/the-garden-of-new-zealand

18 'The garden in literature', *Press*, 10 March 1900.

19 'City reserves: Annual inspection', *Press*, 13 March 1924.

20 Ibid.

21 New Zealand Federation of University Women, Canterbury Branch, *Sydenham: The model borough of old Christchurch: An informal history* (Christchurch: New Zealand Federation of University Women Canterbury Branch, 1977), 85.

22 'Work for the week', *Star*, 20 May 1893.

23 'Chrysanthemum show', *Press*, 3 May 1895.

24 Ebenezer Howard, *Garden Cities of Tomorrow* (London: Faber and Faber, 1945), 48 (Howard's emphasis).

25 Ebenezer Howard, *Garden Cities of Tomorrow* (London: Swan Sonnenschein & Co., 1902), 22–25.

26 Ibid., 13.

27 'Relieving London', *Press*, 27 January 1900.

28 Ibid.

29 W. Feuchtwanger, 'Gorst, Sir John Eldon', in H. Matthew and B. Harrison (eds), *Oxford Dictionary of National Biography: From the earliest times to the year 2000*, vol. 22 (Oxford: Oxford University Press, 2004), 1018.

30 'Jam and pickles', *Press*, 5 March 1904.

31 K.O. Arvidson (ed.), 'Editor's introduction', in *John Eldon Gorst: The Māori king* (Auckland: Reed, 2001, 3rd edn), ix.

32 John Gorst, *The Children of the Nation: How their health and vigour should be promoted by the state* (London: Methuen & Co., 1907), 274–75.

33 Ibid., 276.

34 Ibid., 232–33.

35 'Work for the week', *Star*, 14 July 1906.

36 James Cowan, *The Official Record of the New Zealand International Exhibition of Arts and Industries Held at Christchurch, 1906–7: A descriptive and historical account* (Wellington: Government Printers, 1910), 383.

37 Ibid., 384.

38 Ibid.

39 Edward J. Wakefield, *Adventure in New Zealand from 1839 to 1844*, vols 1 and 2 (London: John Murray, 1845).

40 'Dunedin Horticultural Society', *Otago Witness*, 2 June 1898, 8.

41 'Dunedin Horticultural Society', *Otago Witness*, 26 January 1899, 14.

42 'Milton', *Otago Witness*, 8 October 1902, 35; 'Milton', *Otago Witness*, 17 February 1904, 32.

43 'Kaitangata Horticultural Show', *Otago Witness*, 24 February 1909, 10.

44 'Garden tools', *Evening Post*, 13 October 1900, 4.

45 Tunnicliff (ed.), *The Selected Letters of Mary Hobhouse*, 78; *Evening Post*, 13 October 1900, 4.

46 W. Hyde, 'The Homestead Beautiful', *Journal of Agriculture*, 15 August 1931, 131.

47 First described by French art critic and collector Philippe Burty in 1872, Japonism is the study of Japanese art and artistic talent: https://en.wikipedia.org/wiki/Japonism. See also: J. Beattie, J. Heinzen and J.P. Adam, 'Japanese Gardens in New Zealand, 1850–1950: Transculturation and transmission', *Studies in the History of Gardens and Designed Landscapes*, vol. 28, no. 2, April–June 2008, 219–36.

48 William Turton, Diary, 13 March, 16 March, 28 March 1922, Barbara Musgrave Private Collection.

49 Ibid., 20 January, 30 July, 27 August 1904.

50 Ibid., 15 February, 16 February, 27 March, 17 April, 30 August, 31 August 1922.

51 Ibid., 27 October 1923.

52 Ibid., 1–5 December 1924.

53 *New Zealand Herald*, 3 March 1919, 2.

54 'The garden', 20 February 1919, newspaper unknown, Māori Committee (Waikato) Papers, MSY-50008, ATL.

55 *New Zealand Herald*, 15 March 1919, supplement, 3.

56 *Observer*, 24 July 1909, 18.

57 *Evening Post*, 29 August 1910, 10.

58 *New Zealand Herald*, 2 January 1911, 2.

59 Ibid.

60 Rupert Tipples, *Colonial Landscape Gardener: Alfred Buxton of Christchurch, New Zealand, 1872–1950* (Lincoln: Lincoln University Department of Horticulture and Landscape, 1989), 28.

61 Harry R. Wills, 5 January 1927, 1 October 1927, 22 October 1927, 13 January 1928, box 2, folder 10, item 8, ARC1991.74, CM.

62 Ibid., 6 January 1928.

63 The same apple is referred to by George Gordon's grandson in his recollection of George's Holly Road garden: 'I well remember the large red, almost black, "Black Prince" apple.' New Zealand Federation of University Women, Canterbury Branch, *St Albans: Swamp to suburbs, an informal history* (Christchurch: New Zealand Federation of University Women, Canterbury Branch, St Albans, 1989), 159.

64 Audrey Potter, interviewed 24 August 2005.

65 Ibid.

66 June Stewart, interviewed 1 September 2005.

67 Ibid.

68 Rosemary McLeod, 'In the rose know', *North & South*, April 1991, 8.

69 Gardening Circle Minutes, 17 May 1916, Otago Women's Club: Records, AG-642-06/01, Hocken Collections, Te Uare Taoka o Hākena.

70 *Grey River Argus*, 22 January 1918, 3.

71 Gardening Circle Minutes, 9 July 1917, 13 August 1917, Otago Women's Club: Records, AG-642-06/01, Hocken Collections, Te Uare Taoka o Hākena.

72 '1st Annual Meeting', Wadestown Cottage Garden and Beautifying Society Minute Book 1918–1922, MSX-3512, ATL.

73 Wadestown Cottage Garden and Beautifying Society Minute Book 1918–1922, MSX-3512, ATL.

74 Ibid., 20 February 1919, 30 May 1919, 18 June 1919, 11 September 1919, 20 November 1919, 23 June 1921.

75 C. Amodeo, *Wilderness to Garden City: A celebration of 150 years of horticultural endeavour in Canterbury* (Christchurch: Canterbury Horticultural Society, 2001), 48.

76 CHS Minute Books, 12 October 1917, unnumbered volume (1912–1934), CHS; Christine Dann, *Cottage Gardening in New Zealand* (Wellington: Allen and Unwin, 1990), 1–5; CHS Minute Books, 23 January 1919, unnumbered volume (1912–1934), CHS.

77 Wadestown Cottage Garden and Beautifying Society Minute Book 1918–1922, 6 November 1919, MSX-3512, ATL.

78 'Flower show', *Grey River Argus*, 25 November 1920, 4.

79 'Works and gardens: Beautifying the workers' surroundings', *Press*, 11 February 1918.

80 Timaru Beautifying Association Minute Book 1899–1910, 14 April 1899, MS-2149, ATL; 'First annual report', Timaru Beautifying Association Minute Book 1899–1910, MS-2149, ATL.

81 For example, New Zealand Railways to Gardening Circle, 18 October 1927, 19 October 1927, 31 October 1927, 10 November 1927, Gardening Circle Minutes, Otago Women's Club: Records, AG-642-06/01, Hocken Collections, Te Uare Taoka o Hākena.

82 *CB*, July 1928, 3; 'Annual report', 31 May 1930, CHS Minute Books, unnumbered volume (1912–1934), CHS. World War Two would bring a dip in membership numbers, despite the fact that soldiers on service retained honorary memberships.

83 Ursula Bethell to Frank Sidgwick, 23 July 1932, MB microfilm 96/1, MB.

84 'Gardens of the Bible', *Press*, 5 February 1934.

85 'Gladiolus and dahlia: Horticultural Society's show', *Press*, 2 February 1934.

86 'Horticulture: Commercial travellers' competition', *Press*, 10 February 1934.

87 'Editorial', *LT*, 24 February 1925; see also Amodeo, *Wilderness*, 154–55.

88 Jack Humm, 'Delightful gardens at Sumner and Redcliffs', *CB*, August 1930, 9.

89 Geoffrey Rice, *Heaton Rhodes of Otahuna: The illustrated biography* (Christchurch: Canterbury University Press, 2001), 114, 135.

90 Ibid., 78, 292.

91 Prue Lovell-Smith, interviewed 25 January 2006.

92 Lois Morris, interviewed 31 December 2009.

93 Gareth Winter, email, 29 July 2010.

94 Gareth Winter, email, 28 and 29 July 2010.

95 Donald Denoon and Philippa Mein Smith, *A History of Australia, New Zealand and the Pacific* (Oxford: Blackwell, 2000), 292; Michael King, *The Penguin History of New Zealand* (Auckland: Penguin, 2003), 279.

96 11 March 1941, CHS Minute Books (1940–1946), Canterbury Horticultural Society.

97 'Horticulture a boon, redeems life from drudgery', *Timaru Herald*, 8 February 1946, 4.

98 Ibid.

99 Leeston Church, Plan of Planting at Leeston Parsonage, 1943, Methodist Church of New Zealand Archives.

4 Native plants in a fragile paradise: 1914–35

1 www.kew.org/read-and-watch/imperial-botany-and-networks-of-science

2 Frances Loeffler, 'Pteridomania: A visual history of the fern in New Zealand' (MA thesis, Victoria University of Wellington, 2006), 52.

3 Paul Star and Lynne Lochhead, 'Children of the Burnt Bush: New Zealanders and the indigenous remnant, 1880–1930', in E. Pawson and T. Brooking, *Making a New Land* (Dunedin: Otago University Press, 2013), 154.

4 John Williams, Journal: 1842–1844, NZMS 1063, Auckland Libraries, Sir George Grey Special Collections.

5 Ibid.

6 The Journal of Mrs Henry Williams, 1844–1850, 6 August 1844, NZMS 1044, Auckland Libraries, Sir George Grey Special Collections.

7 'Horticultural society', *New Zealand Gazette and Wellington Spectator*, 31 December 1842, 3; 18 February 1843, 2.

8 'Observations', *Otago Witness*, 16 August 1851.

9 'To botanists and others', *Wellington Independent*, 30 June 1866, 4.

10 Loeffler, 'Pteridomania', 49. The Wardian case was 'an early version of the terrarium. It found great use in the 19th century in protecting foreign plants imported to Europe from overseas, the great majority of which had previously died from exposure during long sea journeys, frustrating the many scientific and amateur botanists of the time': https://en.wikipedia.org/wiki/Wardian_case

11 Loeffler, 'Pteridomania', 50.

12 'Horticultural show', *Press*, 2 March 1866.

13 'Horticultural society's show', *Press*, 1 March 1872.

14 'Horticultural society', *Press*, 7 March 1884; Adams & Sons to Hooker, 15 August 1855, AJCP M744, Reel 15, ffl-18 – Directors' Correspondence vol. 174, New Zealand Letters ADA-HEC, 1854–1900, Royal Botanic Gardens, Kew.

15 Ibid., 6 February 1894.

16 M. Murphy, *Handbook of Gardening for New Zealand with a Section on Poultry and Bee-Keeping* (Christchurch: Whitcombe & Tombs, 3rd edn, 1895), 168–71.

17 Julius von Haast to Hooker, 1 August 1867, Royal Botanic Gardens, Kew, AJCP M745, Reel 16, f334 Directors' Correspondence Volume 174, New Zealand Letters ADA-HEC, 1854–1900.

18 Haast to Hooker, 12 February 1869, AJCP M745, Reel 16, f352 Directors' Correspondence Volume 174, New Zealand Letters ADA-HEC, 1854–1900, Royal Botanic Gardens, Kew.

19 Haast to Hooker, 28 June 1870, AJCP M745, Reel 16, f361 Directors' Correspondence Volume 174, New Zealand Letters ADA-HEC, 1854–1900, Royal Botanic Gardens, Kew.

20 Adams & Sons to Hooker, 7 April 1886, AJCP M744, Reel 15, unnumbered item Directors' Correspondence Volume 174, New Zealand Letters ADA-HEC, 1854–1900, Royal Botanic Gardens, Kew.

21 Frederick Chapman to Arthur Hill, received 18 May 1927, AJCP M740, Reel 11, f208 Directors' Correspondence Volume 169 Australia Letters 1914–1928, Royal Botanic Gardens, Kew.

22 Hill to Leonard Cockayne, 22 June 1927, AJCP M740, Reel 11, f289 Directors' Correspondence Volume 169 Australia Letters 1914–1928, Royal Botanic Gardens, Kew.

23 Cockayne to Hill, 12 August 1927, AJCP M740, Reel 11, f290 Directors' Correspondence Volume 169 Australia Letters 1914–1928, Royal Botanic Gardens, Kew.

24 'Prominent botanist to tour dominion. Dr Hill's visit from Kew', *CB*, January 1928, 1; 'Kew and Empire', *CB*, February 1928, 21; E. Phillips Turner to Joseph Kinsey, 4 February 1928, Kinsey, Sir Joseph James Papers, 1869–1935, Reel 2, ARCH 1991.14, CM.

25 Jonathan West, *The Face of Nature: An environmental history of the Otago Peninsula* (Dunedin: Otago University Press, 2017), 267–68.

26 Star and Lochhead, 'Children of the Burnt Bush', 144.

27 For contemporaneous scholarship see Gordon Jacks and Robert Whyte, *The Rape of the Earth: A world survey of soil erosion* (London: Faber and Faber, 1939); Albert Howard, *An Agricultural Testament* (London: Oxford University Press, 1940), especially chapter 10, 'Some Diseases of the Soil'; H. Massingham (ed.), *England and the Farmer: A symposium* (London: B.T. Batsford, 1941). See also Lance McCaskill, *Hold this Land: A history of soil conservation in New Zealand* (Wellington: A.H. & A.W. Reed, 1973); Gregory Barton, 'Empire forestry and the origins of environmentalism', *Journal of Historical Geography*, vol. 27, no. 4, 2001, 529–52; Gregory Barton, 'Sir Albert Howard and the forestry roots of the organic farming movement', *Agricultural History*, vol. 75, no. 2, 2001, 168–87; James Beattie, 'Environmental anxiety in New Zealand, 1840–1941: Climate change, soil erosion, sand drift, flooding and forest conservation', *Environment & History*, vol. 9, no. 4, 2003, 379–92.

28 'Preservation of native bush', *Press*, 4 June 1900.

29 Christchurch Beautifying Association, Minute Book, 26 February 1906, vol. 1, 11–12; 1 September 1908, vol. 1, 73 and 'Chairman's report', 84, CM.

30 Gordon Ogilvie, *The Port Hills of Christchurch* (Wellington: A.H. & A.W. Reed, 1978), 212.

31 A.H. McLintock, '1913 Royal Commission and the Formation of the State Forest Service', *Te Ara – the Encyclopedia of New Zealand*: www.TeAra.govt.nz/en/1966/forestry?page-5

32 'Royal Commission on Forestry', *AJHR*, 1913, xiv–xv.

33 E.V. Sanderson, *F&B*, no. 31, 1933.

34 Michael Roche, 'The State as Conservationist, 1920–1960: "Wise use" of forests, lands, and water', in Eric Pawson and Tom Brooking (eds), *Environmental Histories of New Zealand* (Melbourne: Oxford University Press, 2002), 191–92.

35 E.V. Sanderson, *F&B*, vol. 1, April 1924.

36 'Mountain daisies· Celmisia', *CB*, January 1928, 17.

37 Samuel Page, 'Our native flora: Species of clematis in the vicinity of Christchurch', *CB*, June 1930, 5.

38 G. Butler, 'Arthur's Pass National Park', *CB*, April 1928, 17.

39 'National Park at Arthur's Pass: A board of control', *CB*, April 1928, 19–21.

40 See, for example, Anna Petersen, 'The European use of Māori art in New Zealand homes c. 1890–1914', in Barbara Brookes (ed.), *At Home in New Zealand: History, houses, people* (Wellington: Bridget Williams Brooks, 2000), 57–72.

41 David Tannock, *Manual of Gardening in New Zealand* (Auckland: Whitcombe and Tombs, 1914), 65, 68. Tannock was instrumental in the development of much of Dunedin Botanic Gardens.

42 Murphy, *Handbook for Gardening in New Zealand*, 168, 171.

43 A. Wilson, 'Alpines', *Otago Witness*, 21 August 1901, 10.

44 James Speden, 'Veronica dasyphylla: One of the choicest gems of our New Zealand native flora', *CB*, May 1929, 26.

45 'Alpine pinks', *Star*, 5 October 1906, 1. Transylvania is a region in Romania. 'Garden notes', *Press*, 15 October 1910.

46 A. Wilkinson to Hill, 29 July 1928, AJCP M740, Reel 11, f347 Directors' Correspondence Volume 169 Australia Letters 1914–1928, Royal Botanic Gardens, Kew; J.T. Sinclair, *Press*, 2 June 1920; W.J. Humm, 'Alpine campanulas: The bell flower', *CB*, April 1929, 9; Sinclair, 'In the garden', *Press*, 8 April 1930; Fawcett Clapperton, 'Autumn in the rock garden', *CB*, March 1930, 7.

47 Leonard Cockayne, *The Cultivation of New Zealand Plants* (Auckland: Whitcombe & Tombs, 1923), 62–63. Noting that veronicas were found in many ecosystems, including 'the great screes of the Southern Alps', Cockayne recommended the 'whipcord' species for use in the alpine garden.

48 'Plants for the home', *NZ Truth*, 21 June 1924, 14.

49 'Honor the native tree: Winter show exhibits', *NZ Truth*, 19 July 1924, 8.

50 Unnamed newspaper clipping regarding lecture of 2 September 1927, in Gardening Circle Minutes, Otago Women's Club: Records, AG-642-06/02, Hocken Collections, Te Uare Taoka o Hākena, University of Otago.

51 A selection of the articles for 1928 demonstrate this point adequately: 'Mountain daisies: Celmisia', *CB*, January 1928, 14–17; 'New Zealand native flora: Two beautiful flowers', *CB*, March 1928, 10–11; C. Jones, 'The rock garden', *CB*, April 1928, 23–24; C. Jones, 'The rock garden', *CB*, May 1928, 20–22; C. Jones, 'The rock garden', *CB*,

June 1928, 21–25; 'Mountain gardens: A botanising trip', *CB*, June 1928, 27–29; J. Speden, 'The flora of Arthur's Pass', *CB*, July 1928, 11–13; C. Jones, 'Plants suitable for the rock garden', *CB*, July 1928, 27–29; A. Wilkinson, 'Alpine and rock gardens', *CB*, August 1928, 32–35; J. Speden, 'The flora of Arthur's Pass', *CB*, September 1928, 12–14; J. Speden, 'The flora of Arthur's Pass', *CB*, October 1928, 27–29; C. Reece, 'Native plants in our gardens', *CB*, October 1928, 31–32; 'New Zealand native veronicas', *CB*, November 1928, 19–21; 'Ranunculus Paucifolius', *CB*, November 1928, 20.

52 'Mountain daisies: Celmisia', *CB*, January 1928, 17.

53 George Fenwick, 'Some plants of New Zealand and the Sub-Antarctic Islands', *CB*, December 1928, 11, 12.

54 Ivory Brothers Collection, B9 E5, MB; J.M. Baxter, *Descriptive Catalogue of New Zealand Native Trees, Shrubs and Plants* (Christchurch, 1928), 21–22. These were Christchurch nurseries.

55 M.J. Barnett, 'The rock garden', *CB*, June 1930, 24.

56 '"Rock gardens" competition', *CB*, November 1932, 3; 'Home gardens: Results of judging at Sumner and Cashmere', *CB*, December 1933, 24; 'Canterbury Horticultural Society: Annual report for the year ending 31 May 1934', *CB*, June 1934, 23; 'Annual report', *CB*, June 1935, 7; 'C.H.S. garden competitions', *CB*, January 1936, 3; 'Canterbury Horticultural Society: Annual report for the year ending 31st May 1936', *CB*, February 1936, 7; 'Canterbury Horticultural Society: Annual report for the year ending 31st May 1937', *CB*, June 1937, 6; 'Gardens competitions', *CB*, February 1946, 5; 'Home garden competition results, 1950', *CB*, March 1950, 4.

57 Given the accompanying illustration, the writing style and garden description, it is assumed that the author of this article is H. McD. Vincent. 'Alpine and rock gardening', *CB*, April 1932, 27.

58 H. McD. Vincent, 'Charm should be the ideal in the alpine garden', *CB*, June 1932, 9.

59 H. McD. Vincent, 'Fancy can run free in the alpine garden', *CB*, August 1932, 21.

60 Ibid., 20.

61 '"Rock gardens" competition', *CB*, November 1932, 4.

62 'Rock gardens competition: Report by the judges', *CB*, December 1932, 15.

63 '"Rock gardens" competition', *CB*, November 1932, 4.

64 'C.H.S. rock gardens competition', *CB*, January 1937, 12.

65 Jean Lawrence, *Gardening Tales* (Wellington: Millward Press, 1981), 19–21.

66 Gardening Circle Minutes, 12 April 1916, 20 October 1915, 17 May 1917, 18 September 1917, Otago Women's Club: Records, AG-642-06/01, Hocken Collections, Te Uare Taoka o Hākena.

67 'Obituary: Miss Margaret Scott Johnston', *Otautau Standard and Wallace County Chronicle*, 19 September 1922, 3.

68 Alexander Paterson to Nelson Rock Garden Society, 11 May 1926, Paterson, Alexander Stronach: Letterbook, Misc-MS-1104, Hocken Collections, Te Uare Taoka o Hākena.

69 Ibid., and 25 August 1926, 6 May 1927.

70 Ibid., 11 May 1926.

71 Kirstie Ross, *Going Bush: New Zealanders and nature in the twentieth century* (Auckland: Auckland University Press, 2008), 11.

72 James Speden, 'The flora of Arthur's Pass', *CB*, October 1928, 29.

73 A. Tyndall, 'The destruction of our native flora', *CB*, June 1929, 12–15; 'City's gardens: Native shrubs popular in Christchurch', *Press*, 1 February 1935.

74 June Stewart, interviewed 1 September 2005.

75 J.A. McPherson, untitled paper, 1932, 'Ferns and ferneries', CH 335, Box 22, Item /23, Christchurch City Council Archives.

76 Kathleen Guy, interviewed 21 November 2005.

77 W.J. Humm, 'Delightful gardens at Sumner and Redcliffs', *CB*, February 1930, 9, 10.

78 Ibid., 8.

79 C.H. Reece, 'Te Kiteroa', *CB*, November 1928, 23.

80 Alfred Buxton (1872–1950) was a landscape gardener and nurseryman. Rupert Tipples, *Colonial Landscape Gardener: Alfred Buxton of Christchurch, New Zealand, 1872–1950* (Lincoln: Lincoln University Department of Horticulture and Landscape, 1989), 93; '"Holmhurst": A charming residence at Cashmere', *CB*, April 1928, 11; John Macmillan Brown, 'Professor Macmillan Brown's garden at Cashmere', *CB*, March 1928, 3–5.

81 'Sumner and Redcliffs garden competition', *CB*, January 1930, 9.

82 Reece, 'Te Kiteroa', 23; *CB*, April 1931, 11, 13.

83 Brown, 'Professor Macmillan Brown's Garden at Cashmere', 6.

84 Kate Jordan, 'Golden weather gardening: New Zealand home sections, 1945–1970' (MA thesis, University of Auckland, 2010), 36.

85 'Annual report', *CB*, June 1946, 6.

86 'News of the society', *CB*, December 1947, 4.

87 'News of the society', *CB*, September 1948, 4; November 1949, 4; October 1951, 4.

88 Winifred Chapman, '"Te Wharekoa": "The house of gladness"', *CB*, March 1930, 3–4.

89 Ibid., 5.

90 Audrey Potter, interviewed 24 August 2005.

91 Judith Todd, interviewed 31 August 2005.

92 Cushla Barker, interviewed 25 August 2005.

93 Warren Duncan, interviewed 19 December 2005.

94 Alison Helm, interviewed 5 December 2005.

95 Fern Every, interviewed 16 July 2005.

96 Hugh England, interviewed 9 February 2006.

97 Kathleen Guy, interviewed 2 September 2005.

98 Ibid.

99 Prue Lovell-Smith, interviewed 25 January 2006.

100 Jennifer Barrer, interviewed 18 November 2005.

101 Florence Harsant, 'Early Life', NZMS 947, 9–10, Auckland Libraries, Sir George Grey Special Collections.

102 Florence Harsant, 'Our Waitahanui garden', NZMS 947, 1, Auckland Libraries, Sir George Grey Special Collections.

103 Ibid., 2.

104 'The garden', 20 February 1919, newspaper unknown, Māori Committee (Waikato) Papers, MSY-50008, ATL.

105 James Belich, *Paradise Reforged: A history of the New Zealanders from the 1880s to the year 2000* (Auckland: Penguin, 2001), 196.

106 Arohanui Lawrence, interviewed 21 January 2010.

107 Ngaio Marsh, *Black Beech and Honeydew* (London: Harper Collins, 2002 edn), 86; Ursula Bethell, 'Primavera', in *From a Garden in the Antipodes* (London: Sidgwick & Jackson, 1929), 19.

108 'Gardens for songsters – cages for criminals! How gardeners may be hosts of sweet-singing guests', *F&B*, May 1936, 11.

109 'Quick success with native trees', *F&B*, February 1936, 5.

110 Lance McCaskill, 'Growth rate of native trees under cultivation', *F&B*, May 1940, 8–10.

111 Wanderer, 'Lost avian paradise', *F&B*, August 1946, 3.

112 Ethel Musgrave, Diary, 27 August 1947, Barbara Musgrave Collection.

113 'Bird and man: Housing competition', *F&B*, February 1947, 3.

114 Hope Gibbons, 'Annual report', in 'Native Plant Preservation Society (Inc) New Zealand, Annual Report and Balance Sheet', 30 June 1938, 8, Alexander Turnbull Library.

115 Lance McCaskill, *Hold This Land: A history of soil conservation in New Zealand* (Wellington: Reed, 1973).

116 'Editorial', *Press*, 20 March 1940, and McCaskill's reply to adults, 'Soil erosion: Investigation and control in Ohio, a lesson for New Zealand', *Press*, 20 March 1940. 'Nature notes: Soil erosion', *Press*, 13 April 1940. McCaskill's 'Nature Notes' usually featured a native plant.

5 Vegetables, compost and the Dig for Victory campaign: 1930–50

1 Philippa Mein Smith, *Mothers and King Baby: Infant survival and welfare in an imperial world: Australia 1880–1950* (London: Macmillan Press, 1997), 25, 30.

2 Ibid., 32.

3 It was also echoed in his own gardening. See James Beattie, 'Scientific Agriculture, Health and Gardening: Japan, New Zealand and Bella and Frederic Truby King', *New Zealand Journal of Asian Studies*, vol. 16, no. 2, December 2014, 47–76.

4 David Combridge, 'Gardening notes', *New Zealand New Health Journal*, February 1927, 5; April–May 1927, 4.

5 'Relief workers' garden', *Press*, 24 July 1934.

6 Miles Fairburn, *The Ideal Society and its Enemies: The foundations of modern New Zealand society, 1850–1900* (Auckland: Auckland University Press, 1989), 100.

7   James Watson, 'Crisis and change: Economic crisis and technological change between the world wars, with special reference to Christchurch, 1926–36' (PhD thesis, University of Canterbury, 1984), 562.

8   Maurice Staunton, interviewed 29 August 2005.

9   Shaw to Reed, 16 July 1933 or 1934, 23 April 1934 or 1935, 18 April 1937, 14 July 1937, 7 July 1938, 22 May 1938, 23 July 1939, 18 August 1940, 3 August 1941, 26 July 1942, 19 August 1943, 6 August 1944, 29 July 1946, Shaw, J.W., NZMS 1500, Auckland Libraries, Sir George Grey Special Collections.

10   Margaret Jones, interviewed 15 June 2009.

11   http://thecommunityarchive.org.nz/node/67115/description. See also Susan Wilson, 'The Aims and Ideology of Cora Wilding and the Sunlight League 1930–1936' (MA essay, University of Canterbury, 1980), 31–32; Nadia Gush, 'Beauty of Health: Cora Wilding and the Sunlight League' (MA thesis, University of Canterbury, 2003).

12   Sunlight League Gardening Group Minutes, 7 June 1941, 26 July 1941, Cora Wilding Papers, MB 183-1.4, *MB*.

13   Ibid., 20 September 1941; January 1944; July 1945.

14   www.nzhistory.govt.nz/women-together/new-zealand-womens-food-value-league

15   *Bulletin of the NZ Women's Food Value League*, vol. 1, no. 1, 1937, 1, ATL.

16   https://nzhistory.govt.nz/women-together/new-zealand-womens-food-value-league

17   C.L. Gillies, 'Social revolution and the garden', *Bulletin of the NZ Women's Food Value League*, vol. 6, no. 1, 1942, 3–4, ATL; Guy Chapman, 'Soil and Nutrition', *Compost Magazine*, vol. 7, no. 6, 1948, 16.

18   Canterbury Association of Commercial Gardeners Incorporated to Town Clerk, 9 June 1942, CH 342, Box 72, Christchurch City Council Archives.

19   Christchurch City Council Minute Book no. 1, CH 380/73, 22883, Christchurch City Council Archives.

20   Christchurch Fruit & Produce Brokers' Association to Town Clerk, 9 June 1942, CH 342, Box, 72, Christchurch City Council Archives.

21   Vegetable Committee Minutes, 1 July 1942, Special Committee Minute Book No. 8, CH 380/c/124, 1746, Christchurch City Council Archives.

22   'His Worship the Mayor's Statement', 28 September 1942, Christchurch City Council Minute Book No. 1, CH 380/73, 22968, Christchurch City Council Archives.

23   Nutritional Value and General Committee, Civic Vegetable Campaign Minutes, 23 February 1943, 3 February 1943, 11 February 1943, 2 March 1943, Special Committee Minute Book No. 8, CH 380/c/124, 1845, Christchurch City Council Archives.

24   Ibid., 16 March 1943.

25   James Belich, *Paradise Reforged: A history of the New Zealanders from the 1880s to the year 2000* (Auckland: Allen Lane, 2001), 289, 291.

26   Nancy Taylor, *The New Zealand People at War: The home front*, vol. 2 (Wellington: Department of Internal Affairs, 1986), 782–83.

27   Ibid., 784.

28   'Community garden at transit camp', *Auckland Star*, 20 November 1944, 4.

29   Taylor, *The New Zealand People at War*, vol. 2, 784, 785.

30 Vegetable Committee, 15 June 1943, Special Committee Minute Book No. 8, CH 380/c/124, 1885, Christchurch City Council Archives.

31 Nutritional Value and General Committee Minutes, 2 March 1943, Special Committee Minute Book No. 8, CH 380/c/124, 1842, Christchurch City Council Archives.

32 Enid Roberts, *Remembered: The life and work of Ben Roberts, M.P.* (Masterton: Masterton Publishing, 1965), 77; Minutes of Inaugural Meeting of Representatives to set up Canterbury Victory Garden Council, 14 September 1944, Special Committee Minute Book No. 8, p.1, CH 380/c/1246, 2006, ANZCRD.

33 Minutes of Inaugral Meeting, 2.

34 Canterbury Victory Garden Council Minutes, 21 September 1944, Ag 93/3/29, Christchurch City Council Archives. (Please note that since the early 1980s, when the Department of Agriculture duplicates used in this book were copied from Archives New Zealand, the classification system has changed. The new online cataloguing system, 'Archway', is still incomplete and the documents in question are currently untraceable. Staff at both the Christchurch and Wellington offices have searched for these documents unsuccessfully. The copies used here are the property of Professor John Cookson. The archives have now been shifted to the Christchurch City Council.)

35 Ibid., 19 October 1944, 2 November 1944, Ag 93/3/29, Christchurch City Council Archives.

36 Audrey Potter, interviewed 24 August 2005.

37 Judith Todd, interviewed 31 August 2005.

38 Dorothy and Wray Fee, interviewed 24 August 2005.

39 Prue Lovell-Smith, interviewed 25 January 2006.

40 Pip Middleton, interviewed 31 August 2005.

41 James Sagar, 'The hidden economy: Vegetable gardening and self-sustenance in Caversham, 1895–1947', 2001, 6, unpublished report, MS-2690/121, Hocken Collections, Te Uare Taoka o Hākena.

42 Andrea Gaynor, *Harvest of the Suburbs: An environmental history of growing food in Australian cities* (Perth: University of Western Australia, 2006), 110, 114–15.

43 Shaw to Reed, 13 November 1938, Shaw, J.W., NZMS 1500, Auckland Libraries, Sir George Grey Special Collections.

44 Shaw to Reed, 3 May 1942.

45 Shaw to Reed, 29 July 1946.

46 Shaw to Reed, 9 August 1948.

47 J.T. Sinclair, 'In the garden', *Press*, 16 June 1920.

48 Sinclair, 'In the garden', *Press*, 30 June 1920, 16 June 1920.

49 Sinclair, 'In the garden', *Press*, 9 January 1934.

50 W.J. Humm, 'In the garden', *Press*, 2 October 1934.

51 Humm, 'In the garden', *Press*, 7 March 1939.

52 Humm, 'In the garden', *Press*, 18 January 1938, 25 January 1938.

53 Sinclair, 'In the garden', *Press*, 24 July 1934.

54   Humm, 'In the garden', *Press*, 8 November 1938; see also 1 March 1938.

55   Humm, 'In the garden', *Press*, 2 January 1940. Hortnap is Horticultural Naphthalene (basically mothballs).

56   Swatter, 'Flies and compost', *Press*, 30 December 1943.

57   Swatter, 'Flies and compost', *Press*, 3 January 1944.

58   W.T. Wainman, 'Flies and compost', *Press*, 6 January 1944.

59   Sir Albert Howard's work was well reported in New Zealand. See, for example, 'Sewage disposal', *New Zealand Herald*, 4 February 1937; 'Soil fertility', *Ellesmere Guardian*, 16 April 1937; 'Soil deficiency', *New Zealand Herald*, 23 May 1938; Agricola, '"Soil Must Live"': Maintaining fertility – importance of humus – role of animal manure', *New Zealand Herald*, 24 June 1938; 'Cattle disease: Startling diet theory', *Manawatu Standard*, 18 December 1945. See also Helen Leach, 'The Twentieth-Century Home Garden', in Eric Pawson and Tom Brooking, *Environmental Histories of New Zealand* (Melbourne: Oxford University Press, 2002), 223. Leach mentions that the 1943 edition of the *Yates Garden Guide* included a reference to humus-building, which was tempered, by 1950, with references to artificial fertilisers, chemical herbicides and other pesticides. She refers to this mention of humus as an 'organic message'. Stating that by 1957 artificial compost activators were promoted over animal manures to show that the 'organic message' was dropped overlooks the point that compost, i.e. humus-building, was still, even then, considered important. It should not be assumed that the compost movement and the organic movement were the same, although the origins of the later organic movement held the same objectives. For the international impact of Sir Albert Howard's work, see Gregory Barton, *The Global History of Organic Farming* (Oxford: Oxford University Press, 2018).

60   B. Sprange, quote in NZHCCCB, Minutes of General Meeting, 9 December 1943, MB 259, MB.

61   'News of the day: Longevity's passion', *Gisborne Herald*, 6 November 1950.

62   NZHCCCB, Minutes of the Executive Committee Meeting, 17 May 1945, MB 259, MB.

63   C.E. St John, 'Compost notes', *Press*, 21 April 1945. This is the only one of these columns not written under the pseudonym.

64   T.S. Waugh, *Home Garden Fertilisers and How to Use Them* (Wellington: Reed, 1945), 12.

65   One good example is W.P. Carman's *Wartime Gardening in NZ* (Wellington: Reed, 1942), 8.

66   D.K. Pritchard, *Vegetable Growing in the Home Garden* (Wellington: Dig for Victory Campaign, 1944), 18.

67   'Compost notes', *Press*, 9 June 1945.

68   Lennie, 'Garden notes', *Press*, 12 May 1945.

69   Ethel Musgrave, Diary, 1, 2 January 1947, 10 January 1947, Barbara Musgrave Collection.

70   Ibid., 4 February 1947, 5 February 1947, 7 March 1947.

71   Ibid., 6 March 1947.

72   Ibid., 23 June 1947, 31 July 1947, 20 August 1947.

73 Ibid., 7 November 1947.

74 William Turton, Diary, 23 November 1904, Barbara Musgrave Collection; Ethel Musgrave, Diary, 31 July 1947, 2 August 1947, Barbara Musgrave Collection.

75 Margaret Jones, interviewed 15 June 2009.

76 Lois Morris, interviewed 31 December 2009.

77 'Notes and Comment', *Compost Magazine*, vol. 6, no. 5, September 1947, 1.

78 Special Meeting of the National Council of the New Zealand Humic Compost Club, October 1949. That it did not was partly due to Dove-Myer Robinson's determination, as president, that the society must not give up until its functions were picked up by government.

79 Lennie, 'Garden notes', *Press*, 8 December 1950.

80 Dorothea Turner, 'Suburban Garden Change', 3YA broadcast, 1950, D2231.41, RNZSA.

81 A.G. Kennelly, *The Home Vegetable Garden* (Wellington: Department of Agriculture, 1953), 38, Christchurch Botanic Gardens Library.

82 Lois Morris, interviewed 31 December 2009.

83 Dove Myer Robinson, 'To Members', 24 November 1952, DMR Papers, Box 4, 100.34 (b), NZMS 822, Auckland Libraries, Sir George Grey Special Collections.

84 Yeo Shand, 'The Crime against the Land' (Christchurch: Blundell Bros Ltd., 1941), 24, CCL.

85 Ibid., 7; G. Jacks and R. Whyte, *The Rape of the Earth: A world survey of soil erosion* (London: Faber and Faber, 1939); 'The proposed River Control Bill and the problem – soil erosion', *F&B*, no. 61, August 1941, 9–10.

86 'Quarterly Newsletter', *F&B*, August 1948, 6; Dove-Myer Robinson maintained correspondence with Forest & Bird during 1947–48. DMR Papers, Box 6, 100.60, NZMS 822, Auckland Libraries, Sir George Grey Special Collections. As he wrote to Sir Albert Howard in 1947, he believed that 'allied' organisations like Forest & Bird and the Food Value League should be jointly convened in a nationally representative body to put pressure on the government. Dove-Myer Robinson to Sir Albert Howard, 14 August 1947, DMR Papers, Box 8, 100.75 (d), NZMS 822, Auckland Libraries, Sir George Grey Special Collections.

87 Shand, 'The Crime against the Land', 28.

88 Ibid., 28.

89 Ibid., 17.

90 Ibid., 5.

91 'Food or Profits?', *Compost Magazine*, July 1942, 3.

92 S. Sagar, 'This Plot of Earth', *Compost Magazine*, vol. 5, no. 1, January/February 1946, 7–8.

93 Brian Low, *Land and People in Christian Order* (Christchurch: Presbyterian Bookroom, 1943), 14–17, 4.

94 E. and P. Beaglehole, *Some Modern Maoris* (Christchurch: Whitcombe & Tombs, 1946), 9–10.

95 Ibid., 114.

96 Ibid., 73.

97 Ibid., 74.

98 Joanna Boileau, *Chinese Market Gardening in Australia and New Zealand: Gardens of prosperity* (London: Palgrave Macmillan, 2017), 249–51. There is significant literature on Māori working in Chinese gardens. See, for example, Lily Lee and Ruth Lam, *Sons of the Soil: Chinese market gardeners in New Zealand* (Pukekohe: Dominion Federation for New Zealand Chinese Commercial Growers, 2012); J. Heine, 'Colonial Anxieties and the Construction of Identities: The employment of Māori women in Chinese market gardens, Auckland, 1929' (MA thesis, University of Waikato, 2006).

99 Ernest and Pearl Beaglehole, *Some Modern Maoris* (Christchurch: Whitcombe and Tombs, 1946), 97.

100 *Tuberculosis and the Māori People*: www.nzonscreen.com/title/ tuberculosis-and-the-Māori-people-1952

101 Wiremu Puke, interviewed 19 January 2012.

102 Ebenezer Howard, *Garden Cities of Tomorrow* (London: Faber and Faber, 1902), 25, 32–33.

103 Albert Howard, *An Agricultural Testament* (London: Oxford University Press, 1943), 115.

104 Dove-Myer Robinson to Albert Howard, 22 September 1947, DMR Papers, Box 8, 100.75 (d), NZMS 822, Auckland Libraries, Sir George Grey Special Collections.

105 Roberts, *Remembered*, preface.

6 The rise of toxins in gardens: 1920–80

1   S. Tunnicliff, (ed.), *The Selected Letters of Mary Hobhouse* (Wellington: Daphne Brasell Associates Press, 1992), 73–74.

2   *1865: Hay's Catalogue of Ornamental Trees and Shrubs* (Auckland: G.T. Chapman, 1865), 49–50.

3   Ibid., 62.

4   Gregory Barton, *The Global History of Organic Farming* (Oxford: Oxford University Press, 2018), 7–9.

5   'An Act to prevent the introduction and to provide for the eradication of diseases affecting orchards and gardens', 17 October 1896: www.enzs.auckland.ac.nz/ docs/1896/1896A45.pdf. An overview of the use of sprays in Nelson orchards is provided in Michael Roche, 'Wilderness to Orchard: The export apple industry in Nelson, New Zealand 1908–1940', *Environment and History*, vol. 9, no. 4, 2003, 435–50.

6   'Dragon flies', *Press*, 22 January 1920, 5.

7   Ibid.

8   J. Sinclair, 'In the garden', *Press*, 23 January 1920, 30 January 1920, 5.

9   J. Sinclair, 'In the garden', *Press*, 30 January 1920, 5.

10 J. Sinclair, 'In the garden', *Press*, 20 March 1920, 9.

11 *New Zealand Herald*, 29 March 1919, Supplement, 3.

12 'Fruit pests: New methods of control', *Press*, 10 April 1920, 13.

13 www.inchem.org/documents/icsc/icsc/eics0765.htm

14 'Calcium arsenate', 11 March 1922, unlabelled newspaper clipping, MSY-5008, ATL.

15 'Copper Acetoarsenite, Chemical Datasheet', National Oceanic and Atmospheric Administration, US Department of Commerce, n.d.: https://cameochemicals.noaa.gov/chemical/2981

16 Francis Peryea, 'Historical use of lead arsenate insecticides, resulting soil contamination and implications for soil remediation': www.natres.psu.ac.th/Link/SoilCongress/bdd/symp25/274-t.pdf

17 Ibid.

18 R. Falconer, 'The home garden', *Te Ao Hou*, December 1955, 46.

19 Mary Stroobant, 'Pesticide or suicide?', *Healthy Life*, June/July 1970, 5.

20 www.mfe.govt.nz/publications/hazardous/risks-former-sheep-dip-sites-nov06/html/page14.html; R. Birdsall, 'Commercial aspect and future possibilities of Rotenone', *Industrial & Engineering Chemistry*, vol. 25, June 1933, 642.

21 'Derris Dust: Unobtainable for garden use', *Canberra Times*, 27 June 1942, 4.

22 T. Lennie, 'Garden notes', *Press*, 6 January 1950, 8; 17 February 1950, 2.

23 Meriel Watts, 'Parkinson's Disease, pesticides and Derris Dust', *Organic NZ*, Jan/Feb 2001: www.organicnz.org/organic-nz-magazine/1138/parkinsons-disease-pesticides-and-derris-dust/

24 Murray Livingstone, 'Gardening with Murray', *Paekakariki Xpressed*, 23 November 2007, 37.

25 J. Sinclair, 'In the garden', *Press*, 8 April 1930, 27 May 1930; Advertisement, 'Garden notes', *Press*, 2 January 1940, 2; Advertisement, 'Garden pests routed', *Press*, 2 January 1940, 2.

26 Warren Duncan, interviewed 19 December 2005; Pip Middleton, interviewed 31 August 2005.

27 Lois Morris, interviewed 31 December 2009.

28 Ibid.

29 J. Sinclair, 'In the garden', *Press*, 3 June 1930, 6.

30 *Yates Garden Guide* (Auckland: Arthur Yates & Co., 1957), 287.

31 Ibid., 289–90.

32 Ibid., 300.

33 J. Humm, 'In the garden', *Press*, 30 January 1940.

34 The issue was discussed by the Compost Club in 1946, and it 'was thought the policy of the Branch should be to *encourage the use of Compost* – leaving the artificials to suffer in comparison', NZHCCCB, Minutes of General Meeting, 25 July 1946. The following month the club's executive decided that 'it would be unwise to adopt an aggressive attitude towards artificials', NZHCCCB, Minutes of Executive Committee Meeting, 8 August 1946. In an open debate a year later, the branch president T.D. Lennie argued strongly in favour of artificial fertilisers. NZHCCCB, Minutes of General Meeting, 28 August 1947, MB 259, MB.

35 T. Lennie, 'Garden notes', *Press*, 3 November 1950.

36 *Star Garden Book* (Dunedin: Evening Star Co., 1955), 139.

37 Ibid., 138.

38 Ibid., 16.

39 'New garden feature for Daily Times', *Otago Daily Times*, 24 July 1952, 8.

40 'Garden with Matthews', *Otago Daily Times*, 5 March 1954, 3.

41 'Garden with Matthews', *Otago Daily Times*, 11 March 1955, 3.

42 'Garden with Matthews', *Otago Daily Times*, 18 March 1955, 3.

43 J. Sinclair, 'In the garden', *Press*, 20 March 1920, 9.

44 Cocky, 'Water for birds', 10 January 1950, 2.

45 Rayna Wootton, 'Wootton home garden 1947–1954', Private Collection.

46 Rayna Wootton, 'Garden book 1953', 28 September 1953, 4 October 1953, Private Collection.

47 Rayna Wootton, 'Year 1954', 25 September 1954, Private Collection.

48 Wootton, 'Garden book 1953', 15 November 1953, 21 November 1953, 22 November 1953.

49 Diane Shannon, interviewed 4 January 2010.

50 'Interest shown in home garden competition', *Evening Star*, 13 June 1953, 1; 'Final judging of city gardens this month', *Evening Star*, 6 February 1954, 2; 'Useful pointers from judges of garden contest', *Evening Star*, 8 April 1954, 4.

51 'Winning garden 1956–1957', in Photograph Album of Shows, Festival Displays & Garden Competitions, Dunedin Horticultural Society Inc: Records, AG-524-05, Hocken Collections, Te Uare Taoka o Hākena, University of Otago.

52 Minutes of the Otago Branch of the New Zealand Organic Compost Society, 1 May 1957, unsorted box, Soil Association of New Zealand, 93-003, Hocken Collections, Te Uare Taoka o Hākena, University of Otago.

53 Minutes of the Otago Branch of the New Zealand Organic Compost Society, 6 November 1957 and 17 October 1962, unsorted box, Soil Association of New Zealand, 93-003, Hocken Collections, Te Uare Taoka o Hākena, University of Otago.

54 'The Compost Society has a fine garden in the Kaikorai Valley', *Otago Daily Times*, 4 April 1957, 7.

55 Minutes of the Otago Branch of the New Zealand Organic Compost Society, 13 February 1963, unsorted box, Soil Association of New Zealand, 93-003, Hocken Collections, Te Uare Taoka o Hākena, University of Otago.

56 Ross Home to the Secretary, Otago Branch of the New Zealand Organic Compost Society, 5 May 1967, unsorted box, Soil Association of New Zealand, 93-003, Hocken Collections, Te Uare Taoka o Hākena, University of Otago.

57 Minutes of the Otago Branch of the New Zealand Organic Compost Society, 2 July 1963, and Monthly Newsheet: No. 2, Dunedin Branch of the New Zealand Organic Compost Society, May 1969, unsorted box, Soil Association of New Zealand, 93-003, Hocken Collections, Te Uare Taoka o Hākena, University of Otago.

58 Minutes of the Otago Branch of the Soil Association of New Zealand, 7 July 1970, unsorted box, Soil Association of New Zealand, 93-003, Hocken Collections, Te Uare Taoka o Hākena, University of Otago; 'Horticultural history described', *The Star*, 16 August 1977.

59 Minutes of the Otago Branch of the Soil Association of New Zealand, 2 May 1978, unsorted box, Soil Association of New Zealand, 93-003, Hocken Collections, Te Uare Taoka o Hākena, University of Otago.

60 Cicely Wylie, *A Garden at My Door* (Christchurch: Caxton, 1964), 11.

61 Ibid., 11. Rachel Carson, *Silent Spring* (Boston: Houghton Mifflin, 1962).

62 Wylie, *A Garden at My Door*, 57.

63 Ibid., 56.

64 Lachy Paterson, email, 12 March 2020.

65 Lachy Paterson, email, 3 March 2020.

66 Lachy Paterson, interviewed 21 December 2009.

67 Richard Tankersley, interviewed 9 January 2010.

68 Ibid.

69 Ethel Musgrave, Diary, 2 April 1946, Barbara Musgrave Collection.

70 David Musgrave, interviewed 30 December 2009.

71 Peter Richardson, interviewed 13 February 2010; Roger White, interviewed 31 August 2010; Darcia Solomon, interviewed 13 February 2010.

72 Ibid.

73 Arohanui Lawrence, interviewed 21 January 2010.

74 Rayna Wootton, interviewed 15 September 2005.

75 Pip Middleton, interviewed 31 August 2005.

76 Brian Morris, interviewed 31 December 2009.

77 Brian Gilberthorpe, interviewed 30 August 2005.

78 'Dow attacks study used to ban 2,4,5-T', *Science News*, vol. 115, no. 11, 17 March 1979, 166.

79 Kenneth Morris, 'Our poisoned planet', *Soil and Health*, vol. 38, no. 4, 1979, 29; Anonymous, 'The Soil Association – and 2,4,5-T', *Soil and Health*, vol. 38, no. 4, 1979, 28.

80 A summary of these studies can be found in Appendix B, 'Low levels of dioxin in residential soils at Paritutu in New Plymouth', in Pattle Delamore Partners, *Dioxin Concentrations in Residential Soil, Paritutu, New Plymouth*, 2002, 13, 15: www.mfe. govt.nz/publications/hazards/taranaki-dioxin-report-sep02

81 'Low levels of dioxin in residential soils at Paritutu in New Plymouth': www.mfe. govt.nz/publications/hazards/taranaki-dioxin-report-sep02

82 'Garden with Matthews', *Otago Daily Times*, 10 November 1967, 16.

83 Ibid.

84 'New Jeyes super strength', Advertisement, *Otago Daily Times*, 17 November 1967, 10.

85 See, for example, N. Defarge, J. Spiroux de Vendomois and G. Seralini, 'Toxicity of formulants and heavy metals in glyphosate-based herbicides and other pesticides', *Toxicology Reports* 5 (2018), 156–63; V. Lozano et al., 'Sex-dependent impact of Roundup on the rat gut microbiome', *Toxicology Reports* 5 (2018), 96–107; R. Mesnage, N. Defarge, J. Spiroux de Vendomois and G. Seralini, 'Potential toxic effects of glyphosate and its commercial formulations below regulatory limits',

Food and Chemical Toxicology 84 (2015), 133–53; P. Clausing, 'Glyphosate and cancer: Authorities systematically breach regulations' (Vienna: Friends of the Earth Austria, 2017); R. Mesnage, C. Benbrook and M. Antoniou, 'Insight into the confusion over surfactant co-formulants in glyphosate-based herbicides', *Food and Chemical Toxicology* 128 (2019), 137–45; International Agency for Research on Cancer, 'Monograph on Glyphosate', in *Some Organophosphate Insecticides and Herbicides*, vol. 112 (Lyon: International Agency for Research on Cancer, 2017).

86 Brian Maunder, 'Vegetable gardens and other domestic sources of soil contamination in residential property', Auckland: Earth Consult Limited, n.d.: www.wasteminz.org.nz/wp-content/uploads/Brian-Maunder-Veg.FINALpdf.pdf

87 'Newsletter', no. 4, 1979, unnumbered folder, NZMS 1238, Auckland City Libraries.

88 M. Lusty, *Press*, 'Gardeners' queries', 16 May 1975, 11; 'Value of ladybirds as pest predators', 23 May 1975, 11; 'Codlin moth control', 30 May 1975, 11.

89 M. Lusty, 'Gardeners' queries', *Press*, 30 May 1975, 11.

90 M. Lusty, 'Work for the month – November', *Press*, 5 November 1976, 8.

91 Dallas Matoe, interviewed 10 April 2010.

92 Ami Kennedy, interviewed 1 January 2010.

93 Walter Ian Hamilton, Gardening Diary, 25 May 1987, 11 August 1987, 24 August 1987, MSX-6827, ATL.

7 The collapse and renewal of home gardening culture: 1960–2020

1 Kate Jordan, 'Golden weather gardening: New Zealand home sections, 1945–1970' (MA thesis, University of Auckland, 2010), 52.

2 *The Illustrated Planting Guide* (Christchurch: Woodland, c. 1961).

3 For example, Venia Noble, 'Fuller life for your blooms', *New Zealand Woman's Weekly*, 12 May 1969, 16–18.

4 Jordan, 'Golden weather gardening', 41.

5 *Press*, 2 January 1965.

6 Angela Wanhalla, interviewed 8 February 2010.

7 Gareth Winter, email, 29 July 2010.

8 Ibid.

9 Ibid.

10 S. Challenger, 'Horticulture in Canterbury: Recent activities and developments', *New Zealand Gardener*, vol. 15, no. 11, 1959, 725.

11 Gareth Winter, email 28 July, 2010.

12 *The Illustrated Planting Guide*, 118.

13 Ibid.

14 Gareth Winter, email, 28 July 2010.

15 J. Matthews, *Matthews on Gardening: A practical approach to the growing of fruits, flowers and vegetables* (Wellington: Reed, 1960), 199–200.

16 James McPherson, *The Complete New Zealand Gardener* (Christchurch: Whitcombe & Tombs, 1968), 12.

17 Brian Morris, interviewed 31 December 2010. Brian is the son of John and Florence Morris of Dunedin; Lois is the daughter of Herbert and Dorothy Tyrell.

18 *Press*, 6 November 1976.

19 Ibid. This point is reinforced in Jordan, 'Golden weather gardening', 33.

20 Gareth Winter, email, 28 July 2010. In 1986 Gareth Winter bought Lansdowne Nursery, established in 1895, and owned it until 1997. It was 'exclusively a bedding plant nursery' – emphasising again where popular gardening tastes lay at the time.

21 Diana Musgrave, pers. comm., 4 April 2012.

22 Dallas Matoe, interviewed 10 April 2010.

23 Angela Wanhalla, interviewed 21 December 2009.

24 Tim Shadbolt, 'A personal comment on the urban environment in New Zealand', *The Landscape: Journal of the New Zealand Institute of Landscape Architects Inc.*, no. 21, April 1984, 9.

25 Joanna Pearsall and Bryan Innes, 'Natural swimming pool', *Natural In-Formation*, no. 1, Spring 2000, 1.

26 *Natural In-Formation*, no. 1, Spring 2000, 7.

27 Lachy Paterson, interviewed 21 December 2009.

28 Ibid.

29 Dallas Matoe, interviewed 10 April 2010.

30 Ibid.

31 '21st Anniversary Issue' (Wellington: Webling and Stewart, 1973), Box: 'Seed Catalogues', Crowder Collection.

32 'Yates New Zealand' (Auckland: Arthur Yates & Co, c. 1984), Box: 'Seed Catalogues', Crowder Collection.

33 Angela Wanhalla, interviewed 8 February 2010.

34 Ibid.

35 Lachy Paterson, interviewed 21 December 2009.

36 Jolyon White, interviewed 12 February 2010.

37 Christopher Musgrave, interviewed 28 December 2009.

38 Perhaps there is an interesting history of the home cultivation of drugs to be told one day.

39 Advertisement, *Soil and Health*, vol. 36, no. 4, 1977, 5.

40 Jack Meechin, 'A composting aid for the gardener', *Soil and Health*, vol. 36, no. 4, 1977, 11.

41 George Hutchinson, 'Mechanical composting', *Soil and Health*, vol. 45, no. 4, 1986, 30–31.

42 https://en.wikipedia.org/wiki/Companion_planting

43 Tim Maples, 'Cultural control of pests and diseases', *Soil and Health*, vol. 35, no. 6, 1980, 8.

44 Anonymous, 'Companion planting', *Soil and Health*, vol. 42, no. 4, 1983, 38.

45 Maples, 'Cultural control of pests and diseases'.

46   Soil Association, 'Working with weeds', *Soil and Health*, vol. 46, no. 4, 1987, 20.

47   Robert Crowder, 'IFOAM 94', Box: 'Organic Certification', unnumbered folder, Crowder Collection.

48   Ibid.

49   Robert Crowder to Commander Richard Aylard (Private Sec. to HRH Prince of Wales), 14 October 1994, Box: 'Organic Certification', folder: 'IFOAM 94 Contacts', Crowder Collection.

50   'Constitution of the Doubleday Research Association of New Zealand Incorporated', unnumbered folder NZMS 1238, Auckland Libraries, Sir George Grey Special Collections.

51   'Newsletter', no. 1, August 1977, unnumbered folder NZMS 1238, Auckland Libraries, Sir George Grey Special Collections.

52   *Soil and Health*, February/March 1979, 40; *Soil and Health*, vol. 45, no. 2, 1986, 48.

53   George Maslin, 'The church's story: A guide to the need for social reform' (1984), quoted in Ruth Greenaway, *Living Seasons* (Ōamaru: Maslin/Eaton/Parker Families, 2004), 58.

54   'Tree scheme seen as madness', *Press*, 16 May 1975.

55   Richard Tankersley, interviewed 9 January 2010.

56   Roger Mann, 'Your Garden', in Geoff Bryant (ed.), *The Ultimate New Zealand Gardening Book* (North Shore City: David Bateman, 1995), 13.

57   Diana Anthony, *Creative Sustainable Gardening in New Zealand* (Auckland: David Bateman, 2000).

58   Peggy Kelly, interviewed 3 July 2009.

59   Charles Barrie, interviewed 21 July 2009.

60   Clowance Nolan, 'Native plants in Lincoln township: Ecological baseline survey of residential properties', Lincoln University, 7: www.lincolnenvirotown.org.nz/docs/Native-Plants-in-Lincoln-Township.pdf

61   Gareth Winter, email, 28 July 2010.

62   Bee Dawson, *A History of Gardening in New Zealand* (Auckland: Random House, 2010), 281.

63   Gareth Winter, email, 28 July 2010.

64   Matt Morris, 'A History of Christchurch Home Gardening' (PhD thesis, University of Canterbury, 2006), 166, 189.

65   www.facebook.com/pg/Peterborough-Housing-Co-Op-162357318848/about/; Bill Sykes, interviewed 3 July 2009.

66   John McCrone, 'Return of the commune', *Press*, 25 July 2009, D1. According to Bill Sykes, when the press referred to them as an 'anarchist group' they 'all thought this was great!' Bill Sykes, interviewed 3 July 2009.

67   Ami Kennedy, interviewed 1 January 2010. Ami is Ali Kennedy's daughter.

68   Charles Barrie, interviewed 21 July 2009.

69   David Lang, 'Kelmarna organic gardens', *Doubleday News*, August 1983, 6, unnumbered folder, NZMS 1238, Auckland City Libraries; Paul Lagerstedt to Ian Hamilton, 12 July 1985, Papers Relating to the Devonport Compost Scheme, MS-Papers-5597-18, ATL.

70   'A Devonport organic garden', Papers Relating to the Devonport Compost Scheme, MS-Papers-5597-18, ATL.

71   Golden Bay Sustainable Living Centre: www.gbslc.org/history

72   Jane Kelsey, *The New Zealand Experiment* (Auckland: Auckland University Press and Bridget Williams Books, 1995), 292.

73   Mike Moore, *Children of the Poor* (Christchurch: Canterbury University Press, 1996), 19.

74   Whakahuihui Vercoe, 'A Self-sufficient Māoridom', in Witi Ihimaera (ed.), *Vision Aotearoa: Kaupapa New Zealand* (Wellington: Bridget Williams Books, 1994), 111.

75   'Field trip report: Community gardens tour', *Canterbury Commercial Organics Group Newsletter*, April 2001.

76   Report on the Hagley/Ferrymead Community Board to the Council of 30 January 2002, 28 February 2002: www.ccc.govnz/Council/Proceedings/2002/February/HagleyFerrymead/

77   'St Anne's Community Garden', c. 1998, John Poppleton Collection.

78   John Poppleton, interviewed 8 June 2009. 'Sharing kai and sharing knowledge at "Te Maara @ Cornwall"', c. 2008, John Poppleton Collection.

79   John Poppleton, interviewed 8 June 2009.

80   Charles Barrie, interviewed 21 July 2009.

81   Shadbolt, 'A personal comment on the urban environment in New Zealand'. Shadbolt wrote that army trucks drove 'up and down the 2 acres of kumaras [Ngāti Whātua] had planted'.

82   'Locals stake claim to Edgeware pool land', *Nor'West News*, 26 September 2007, 14.

83   St Albans Working Party, 'Final report to council and community', 13 January 2009, 8, Matt Morris private collection.

84   Ibid., 43.

85   Glenn Brown, interviewed 26 June 2011.

86   Charles Barrie, interviewed 21 July 2009.

87   Megan Blakie, 'Linwood community garden', *Anglican Care*, June/July 2009, 14–15.

88   Ray Wright and Jane Quigley, 'Community and Home Gardens', *Organic Matters*, July 1998, 2, 3.

89   'Organic Groups', *Organic Matters*, August 2000, insert.

90   Minutes of the Christchurch Community Gardens Association, 19 May 2009; the Canterbury Community Gardens Association website listed 29 community gardens in 2019, 24 of which were in Christchurch: www.ccga.org.nz/garden-directory/

91   Matt Morris, 'Local and Community Gardens', in *New Zealand Organic Sector Market Report* (Organics Aotearoa New Zealand, 2018), 53.

92   Ibid., 53.

93   Georgia O'Connor-Harding, 'Group thanked for helping garden recover from thefts', *Pegasus Post*, 4 December 2018.

94   Margaret Jones, interviewed 15 June 2009.

95 Auckland City Council, 'Weed management policy': www.aucklandcity.govt.nz/council/documents/resources/weed/section3-5.asp

96 Mark Hill, 'Christchurch: Organic garden city', *Soil and Health*, vol. 56, no. 3, 1997, 20.

97 'Workshop pushing for "organic city"', *Press*, 8 April 1997, 4.

98 Mark Hill, 'Christchurch: Organic garden city', *Soil and Health*, vol. 56, no. 3, 1997, 22.

99 Frankie Dean, quoted in 'Pupils learn to grow vegetables', *Press*, 3 August 1998.

100 Ami Kennedy, interviewed 1 January 2010.

101 Rachel Sykes to Valmai Becker, 29 May 2001, Organic Garden City Trust Archive.

102 Tāhuri Whenua newsletter, no. 11, Spring 2009.

103 www.gardentotable.org.nz/about

104 Te Ora Hou Ōtautahi, 'Proposal for Windermere site community garden', 2008, Matt Morris private collection.

105 Dallas Matoe, interviewed 10 April 2010.

106 www.transitiontowns.org.nz

107 www.transitiontowns.org.nz/node/341

108 www.transitiontowns.org.nz/node/427

109 'Transition initiative updates', *Organic Matters*, August 2009, 6.

110 Guy Williams, 'Pruning workshop this weekend', *Timaru Courier*, 16 July 2009, 53; www.transitiontowns.org.nz/oamaru

111 Jolyon White, 'Just gardening' (Wellington: Social Justice Commission of the Anglican Church in Aotearoa, New Zealand and Polynesia, 2009), 5.

112 Ibid., 19.

113 Ibid., 21.

114 'About the localising food project': www.localisingfood.com/index.php/the-project

115 Ibid.

116 *Edible Paradise* (2018), director Rich Humphreys.

117 On this see Catherine Bukowski and John Munsell, *The Community Food Forest Handbook: How to plan, organize, and nurture edible gathering places* (White River Junction: Chelsea Green, 2018).

118 http://foodforest.co.nz/about/

119 Robert Guyton, interviewed 19 March 2020.

120 Hayden Munro, 'Backyard veges a growing trend', *Pegasus Bay News*, 29 October 2008, 3; Maggie Barry, 'Natural-born gardeners', *New Zealand Listener*, 15 November 2008.

121 'Beat recession with homegrown fruit', *Star*, 3 July 2009; Lynda Hallinan, 'Plant a tree', *Homegrown: New Zealand Gardener special collector's edition*, 2009, 7.

122 Statistics New Zealand, 'Food Price Index: October 2008', 5: www.stats.govt.nz/products-and-services/hot-off-the-press/food-price-index/food-price-index-oct08-hotp.htm

123 Richard Tankersley, interviewed 9 January 2010. Richard is the son of Robert Tankersley.

124 Sarah Mankelow, 'Sustained by plants', *Press*, 5 August 2008. 'The help, knowledge and aroha of the people of Te Hapū o Ngāti Wheke (Rapaki) was instrumental in the garden's creation': www.doc.govt.nz/our-work/motukarara-conservation-nursery/nga-tipu-whakaoranga-o-tutekawa/

125 Matt Morris, 'A Celestial Place: Hill gardening in a colonial garden city', in *Thesis Eleven*, no. 92, February 2008, 78.

126 Christopher Moore, 'Taking Flight', *Press*, 21 April 2004.

127 Jennifer Barrer, interviewed 27 September 2006.

128 Ibid.

129 Charles Barrie, interviewed 21 July 2009.

130 Darcia Solomon, interviewed 13 February 2010.

131 Roger White, interviewed 31 August 2010.

132 Ibid.

133 Ibid.

134 Ibid.

135 Wiremu Puke, interviewed 19 January 2012; https://hamiltongardens.co.nz/collections/productive-collection/te-parapara-garden/

136 Ibid.

137 Arohanui Lawrence, interviewed 21 January 2010. Arohanui is the daughter of Taanga and Elva Tomoana.

138 Ibid.

139 Hema Wihongi, 'Reintroduction of early kumara varieties' (Auckland: Puhaorangi Māori Culture & Resource Centre, 2010).

140 Arohanui Lawrence, written observations, 22 January 2010, Arohanui Lawrence private collection.

8 Dreams

1 Brendan Hoare, 'Regional food economy: Culture background paper', report for Waitākere City Council, 2009, 10; https://en.wikipedia.org/wiki/Food_sovereignty

2 J. Sagar, 'The hidden economy: Vegetable gardening and self sustenance in Caversham, 1895–1947', unpublished report, 2001, Hocken Collections, Te Uare Taoka o Hākena, University of Otago.

3 Derek Dow, *Māori Health & Government Policy, 1840–1940* (Wellington: Victoria University Press, 1999), 142.

4 Susan Bulmer, 'Ngā māra: Traditional Māori gardens', in M. Bradbury (ed.), *A History of the Garden in New Zealand* (Auckland: Penguin, 1995).

5 Doug Sellman, 'It's time to mind our own business', *Press*, 18 April 2011.

6 James Beattie, 'Science, religion, and drought: Rainmaking experiments and prayers in North Otago, 1889–1911', in J. Beattie, E. O'Gorman and M. Henry (eds), *Climate, Science, and Colonization. Histories from Australia and New Zealand* (New York: Palgrave Macmillan, 2014).

7   The Community Gardening and Food Resilience group consisted of members of Canterbury Community Gardens Association and St Albans Gardening Group.

8   '2015/16 Food Resilience Network Annual Report' (Christchurch: Food Resilience Network, 2016), 6.

9   'Edible Canterbury Charter' (Christchurch: Food Resilience Network, 2015): www.ttophs.govt.nz/vdb/document/1469

10   'Food Resilience Policy' (Christchurch City Council, 2014): www.ccc.govt.nz/the-council/plans-strategies-policies-and-bylaws/policies/sustainability-policies/food-resilience-policy/

11   David Boarder Giles, 'Fear of a Free Lunch: Markets, publics and the would-be gift', in Alison Dundon and Richard Vokes (eds), *Shifting States: New perspectives on security, infrastructure and political affect* (London: Bloomsbury, forthcoming).

12   For work on this in the New Zealand context, see Barton Acres, 'Opportunities for Food Systems Planning in New Zealand' (MPlan thesis, University of Otago, 2010).

13   'Climate Smart Strategy 2012–2025' (Christchurch City Council, 2010): www.ccc.govt.nz/thecouncil/policiesreportsstrategies/strategies/healthyenvironmentstrategies/climatesmart/index.aspx

Bibliography

Abbreviations

ACL Auckland City Libraries
AJHR Appendix to the Journals of the House of Representatives
ANZCRO Archives New Zealand, Christchurch Regional Office
ATL Alexander Turnbull Library, Wellington
BWC British World Conference
CB The City Beautiful
CCL Christchurch City Libraries
CGC Cashmere Garden Club
CGCA Christchurch Guardian and Canterbury Advertiser
CHS Canterbury Horticultural Society
CSS Christchurch Star-Sun
CM Canterbury Museum, Documentary Centre
DMR Dove-Myer Robinson
DO Garden and Landscape Studies Library, Dumbarton Oaks, Washington DC
F&B Forest & Bird
GC Gardeners' Chronicle
GCGI Gardeners' Chronicle and Gardening Illustrated
HG House & Garden
ICHG International Conference of Historical Geographers
LINZ Land Information New Zealand, Christchurch Office
LT Lyttelton Times
MB Macmillan Brown Library, University of Canterbury
NZHCCCB New Zealand Humic Compost Club, Canterbury Branch
RNZSA Radio New Zealand Sound Archives, Christchurch
RW Rayna Wootton Private Collection
SLV State Library of Victoria, Melbourne

Archives

Archives New Zealand/Te Rua Mahara o te Kāwanatanga, Christchurch Regional Office

CH287, ICPS 2692/1864; R22196138 Doctor Donald (Resident Magistrate Lyttelton) to Provincial Secretary – Europeans thieving from native gardens at Rapaki – 7/12/1864

Alexander Turnbull Library

MS Papers 4410, Folder 1, Alex Dickson & Sons to Vereker-Bindon
MS-2149, Timaru Beautifying Association Minute Book 1899–1910
MS-2163, William Trotter, Gardening Diary
MS-2190, William Tye, Diary, 1838–1848

MS-Papers-5597-17, Papers Relating to Devonport Composting Project
MS-Papers-5597-18, Papers Relating to the Devonport Compost Scheme
MS-Papers-7938-48, Walter Ian Hamilton, Gardening Notebook and Related Papers
MS-Papers-8809, Scrapbook
MSX-3424, W.H. Vereker-Bindon collection
MSX-3512, Wadestown Cottage Garden and Beautifying Society Minute Book 1918–1922
MSX-6827, Walter Ian Hamilton, Gardening Diary
MSY-4832, Station Gardener's Diary
MSY-5008, 'Calcium Arsenate', 11 March 1922, unlabelled newspaper clipping
Hope Gibbons, 'Annual Report', in Native Plant Preservation Society (Inc) New
    Zealand, Annual Report and Balance Sheet

Auckland Libraries, Sir George Grey Special Collections

GNZ 635, Auckland Horticultural Society
NZMS 1044, the Journal of Mrs Henry Williams, 1844–1850
NZMS 1063, John Williams, Journal: 1842–1844
NZMS 1238, Doubleday Research Association of New Zealand Incorporated
NZMS 1256, Alice Porter to Will Porter, Correspondence
NZMS 1500, Shaw, J.W.
NZMS 822, DMR Papers, Box 8, 100.75 (d)
NZMS 822, DMR Papers, Box 4, 100.34 (b)
NZMS 865/1, Frederick Spencer, 'Reminiscences of an Old New Zealander'
NZMS 947, Harsant, Florence, Early Life
NZMS 947, Harsant, Florence, Our Waitahanui Garden
NZMS822, DMR Papers, Box 6, 100.60

Australia Joint Copying Project, Royal Botanic Gardens, Kew

M740, Reel 11
M744, Reel 15
M745, Reel 16

Canterbury Community Garden Association

Minutes, 19 May 2009

Canterbury Horticultural Society

Minute Books (1940–1946)
Minute Books (1912–1934)
Minute Books (1903–1908)

Canterbury Museum

Christchurch Beautifying Association Minute Books
475/50, J.R. Godley and E.G. Wakefield Correspondence
ARC 1990.285, Folder 1382, Edward Gibbon Wakefield to Catherine Wakefield, 24
    March 1853
ARCH 1991.14, Sir Joseph James Papers, 1869-1935, Reel 2
ARC1991.74, Harry R. Wills
W.A. Taylor Papers

Cashmere Garden Club Collection
'Margaret Jane (Jean) Foweraker'
Minute Book, 1944–1948

Christchurch City Council Archives
Ag 93/3/29, Canterbury Victory Garden Council Minutes, 21 September 1944
CH 342, /72 Miscellaneous Letters, Inwards, 1939–1944
CH 335, 22/23 Ferns and Ferneries, 1932
CH 380, /73 Christchurch City Council Minute Book
CH 380, c/124 Special Committee Minute Book No.8

Christchurch City Libraries
ARCH 205, Charles Bridge, Diary 1850–1865
ARCH 486, J. Stanley Monck, Diary
Māori Land Court Ikaroa District South Island Minute Book

Food Resilience Network
2015/16 Food Resilience Network Annual Report
Edible Canterbury Charter

Hocken Collections, Te Uare Taoka o Hākena, University of Otago
Misc-MS-0933/002, Belsham, Ulva L., Mrs: Papers
AG-524-05, Dunedin Horticultural Society Inc.: Records
93-003, Soil Association of New Zealand
AG-642-06/01, Otago Women's Club: Records
AG-642-06/02, Otago Women's Club: Records
Misc-MS-0607, Stuart, Alexander: Diary and Recipe Book
Misc-MS-1104, Paterson, Alexander Stronach: Letterbook
MS-2690/121, Sagar, J. 'The hidden economy: Vegetable gardening and self sustenance in Caversham, 1895–1947', unpublished report, 2001

Macmillan Brown Library (MB)
MB 183-1.4, Cora Wilding Papers
MB 259, New Zealand Humic Compost Club Minute Books
MB 512/4 and 512/5, Margaret Elphick Papers
MB microfilm 96/1, Ursula Bethell Papers
Ivory Brothers Collection

Methodist Church of New Zealand Archives
8171 Leeston Church
Folder 43, Wesleyan Missionary Society

Organic Garden City Trust Archive
Kids' Edible Gardens Correspondence

Radio New Zealand Sound Archives
Combridge, D. 'Making a Compost Heap', 3YA broadcast, c. 1947, D6997
Turner, D. 'Suburban Garden Change', 3YA broadcast, 1950, D2231.41

Private Collections

Arohanui Lawrence Private Collection
'Fishing by the moon'

Barbara Musgrave Private Collection
Ethel Musgrave, Diary
William Turton, Diary

Bob Crowder Private Collection
Box: 'Seed Catalogues'
Box: 'Organic Certification', folder: 'IFOAM 94 Contacts'

Brendan Hoare Private Collection
'Regional food economy: Culture background paper', report for Waitākere City
    Council, 2009

John Poppleton Collection
'Sharing Kai and Sharing Knowledge at "Te Maara @ Cornwall"', c. 2008
'St Anne's Community Garden', c. 1998

Rayna Wootton Private Collection
Garden Book 1953
Wootton Home Garden 1947–1954
Year 1954

Te Ora Hou
Proposal for Windermere Site Community Garden, 2008

Interviews

Cushla Barker, interviewed 25 August 2005
Jennifer Barrer, interviewed 18 November 2005
Charles Barrie, interviewed 21 July 2009
Glenn Brown, interviewed 26 June 2011
Warren Duncan, interviewed 19 December 2005
Hugh England, interviewed 9 February 2006
Fern Every, interviewed 16 July 2005
Dorothy and Wray Fee, interviewed 24 August 2005
Brian Gilberthorpe, interviewed 30 August 2005
Kathleen Guy, interviewed 2 September 2005 and 21 November 2005
Robert Guyton, interviewed 19 March 2020
Alison Helm, interviewed 5 December 2005
Margaret Jones, interviewed 15 June 2009
Ami Kennedy, interviewed 1 January 2010
Joan Lamby, interviewed 15 February 2006
Arohanui Lawrence, interviewed 21 January 2010
Prue Lovell-Smith, interviewed 25 January 2006
Dallas Matoe, interviewed 10 April 2010
Pip Middleton, interviewed 31 August 2005

Brian Morris, interviewed 31 December 2009
Lois Morris, interviewed 31 December 2009
Christopher Musgrave, interviewed 28 December 2009
David Musgrave, interviewed 30 December 2009
Lachy Paterson, interviewed 21 December 2009
John Poppleton, interviewed 8 June 2009
Audrey Potter, interviewed 24 August 2005
Wiremu Puke, interviewed 19 January 2012
Peter Richardson, interviewed 13 February 2010
Terry Ryan, interviewed 25 September 2004
Diane Shannon, interviewed 4 January 2010
Darcia Solomon, interviewed 13 February 2010
Maurice Staunton, interviewed 29 August 2005
June Stewart, interviewed 1 September 2005
Bill Sykes, interviewed 3 July 2009
Richard Tankersley, interviewed 9 January 2010
Judith Todd, interviewed 31 August 2005
Angela Wanhalla, interviewed 21 December 2009 and 8 February 2010
Jolyon White, interviewed 12 February 2010
Roger White, interviewed 31 August 2010
Ray Wootton, interviewed 15 September 2005

Periodicals, Newspapers and Official Publications

Anglican Care
Appendix to the Journals of the House of Representatives
Auckland Star
Bulletin of the New Zealand Women's Food Value League
Canberra Times
Canterbury Commercial Organics Group Newsletter
Chip'n'Away
Christchurch Guardian and Canterbury Advertiser
City Beautiful
Compost Magazine
Daily Southern Cross
Employment Matters
Evening Post
Evening Star
Forest & Bird
Grey River Argus
Hawkes Bay Herald
Healthy Life
Journal of Agriculture
Lyttelton Times
Natural In-Formation
New Zealand Colonist and Port Nicholson Advertiser
New Zealand Gazette and Wellington Spectator

New Zealand Herald
New Zealand New Health Journal
New Zealand Woman's Weekly
Nor'West News
North & South
NZ Truth
Observer
Organic Matters
Otago Daily Times
Otago Witness
Otautau Standard and Wallace County Chronicle
Paekakariki Xpressed
Press
Soil and Health
Southern Seed Exchange Newsletter
Star
Tāhuri Whenua Newsletter
Te Ao Hou
Timaru Courier
Timaru Herald
Wellington Independent

Books

'*Star*' *Garden Guide* (Dunedin: The Evening Star, 1955)

1865: Hay's catalogue of ornamental trees and shrubs (Auckland: G.T. Chapman, 1865)

Amodeo, C., *Wilderness to Garden City: A celebration of 150 years of horticultural endeavour in Canterbury* (Christchurch: Canterbury Horticultural Society, 2001)

Andersen, J., *Old Christchurch in Picture and Story* (Christchurch: Simpson and Williams, 1949)

Anderson, A., *The Welcome of Strangers: An ethnohistory of Southern Maori A.D. 1650–1850* (Dunedin: University of Otago Press, 1998)

Anthony, D., *Creative Sustainable Gardening in New Zealand* (Auckland: David Bateman, 2000)

Barton, G., *The Global History of Organic Farming* (Oxford: Oxford University Press, 2018)

Baxter, J.M., *Descriptive Catalogue of New Zealand Native Trees, Shrubs and Plants* (Christchurch, 1928)

Beaglehole, E. and P., *Some Modern Maoris* (Christchurch: Whitcombe & Tombs, 1946)

Beattie, J., *Empire and Environmental Anxiety: Health, science, art and conservation in South Asia and Australasia, 1800–1920* (New York: Palgrave Macmillan, 2011)

Belich, J., *Paradise Reforged: A history of the New Zealanders from the 1880s to the year 2000* (Auckland: Allan Lane, 2001)

Belich, J., *The New Zealand Wars and the Victorian Interpretation of Racial Conflict* (Auckland: Auckland University Press, 1998 edn)

Best, E., *Maori Agriculture* (Wellington: A.R. Shearer, 1976)

Best, E., *The Astronomical Knowledge of the Maori: Genuine and empirical* (Christchurch: Kiwi Publishers, 2002)

Boileau, J. *Chinese Market Gardening in Australia and New Zealand: Gardens of prosperity* (London: Palgrave Macmillan, 2017)

Brockie, W., *New Zealand Alpines in Field and Garden* (Christchurch: Caxton, 1945)

Brookes, B., Cooper, A. and Law, R. (eds), *Sites of Gender: Women, men & modernity in Southern Dunedin, 1890–1939* (Auckland: Auckland University Press, 2003)

Bukowski, C., and Munsell, J., *The Community Food Forest Handbook: How to plan, organize, and nurture edible gathering places* (White River Junction: Chelsea Green, 2018)

Carman, W., *Wartime Gardening in NZ* (Wellington: Reed, 1942)

Chapman, G., *Health Grows Naturally* (Auckland: Democracy Publishing, 1944)

Cockayne, L., *The Cultivation of New Zealand Plants* (Auckland: Whitcombe and Tombs, 1923)

Cruise, Richard, *Journal of a Ten Months' Residence in New Zealand* (1820) (Christchurch: Pegasus Press, 1957 edn)

D'Urville, D., *New Zealand 1826–1827* (trans. Olive Wright) (Wellington: Wingate, 1950)

Dann, D., *Cottage Gardening in New Zealand* (Wellington: Allen and Unwin, 1990)

Davison, G., *The Rise and Fall of Marvellous Melbourne* (Melbourne: Melbourne University Press, 1978)

Dawson, B., *A History of Gardening in New Zealand* (Auckland: Random House, 2010)

Deans, J. (ed.), *Pioneers of Canterbury: Deans letters 1840–1854* (Christchurch: Cadsonbury, 1997)

Denoon, D., and Mein Smith, P., *A History of Australia, New Zealand and the Pacific* (Oxford: Blackwell, 2000)

Dieffenbach, E., *Travels in New Zealand* (London: John Murray, 1843)

Dow, D., *Māori Health & Government Policy, 1840–1940* (Wellington: Victoria University Press, 1999)

Earle, Augustus, *A Narrative of a Nine Months' Residence in New Zealand in 1827* (Christchurch: Whitcombe & Tombs, 1909), 75

Elder, J. (ed.), *The Letters and Journals of Samuel Marsden 1765–1838* (Dunedin: Coulls Somerville Wilkie and A.H. Reed, 1932)

Evison, H., *Te Wai Pounamu: The Greenstone Island: A History of the Southern Maori during the European Colonisation of New Zealand* (Wellington: Aoraki Press, 1993)

Evison, H., *The Long Dispute: Māori land rights and European colonisation in southern New Zealand* (Christchurch: Canterbury University Press, 1997)

Fairburn, M., *The Ideal Society and its Enemies: The foundations of modern New Zealand society 1850–1900* (Auckland: Auckland University Press, 1989)

Frame, J., *Living in the Maniototo* (Auckland: Vintage, 2006)

Gaynor, A., *Harvest of the Suburbs: An environmental history of growing food in Australian cities* (Perth: University of Western Australia, 2006)

Godley, C., *Letters from Early New Zealand by Charlotte Godley 1850–1853* (Christchurch: Whitcombe & Tombs, 1951)

Gorst, J., *The Children of the Nation: How their health and vigour should be promoted by the state* (London: Methuen & Co., 1907, 2nd edn)

Gorst, J., *The Māori King* (Auckland: Reed, 2001)

Greenaway, R., *Living Seasons* (Ōamaru: Maslin/Eaton/Parker Families, 2004)

Grove, R., *Green Imperialism: Colonial expansion, tropical island Edens and the origins of environmentalism* (Cambridge: Cambridge University Press, 1995)

Hallinan, L., 'Plant a tree', in *Homegrown. New Zealand Gardener Special Collector's Edition*, 2009

Heaphy, Charles, *Narrative of a Residence in Various Parts of New Zealand* (London: Smith, Elder & Co., 1842)

Horsman, J., *The Coming of the Pakeha to Auckland Province* (Wellington: Hicks Smith and Sons, 1971)

Howard, A., *An Agricultural Testament* (London: Oxford University Press, 1940, 1943)

Howard, E., *Garden Cities of Tomorrow* (London: Faber and Faber, 1945)

Howard, E., *Garden Cities of Tomorrow* (London: Swan Sonnenschein, 1902)

Hulme, K., *The Bone People* (Auckland: SPIRAL and Hodder and Stoughton, 1985)

Hunt, J., and Wolschke-Bulmahn, J. (eds), *The Vernacular Garden* (Washington, DC: Dumbarton Oaks, 1993)

The Illustrated Planting Guide (Christchurch: Woodland, c. 1960)

Jacks, G., and Whyte, R., *The Rape of the Earth: A world survey of soil erosion* (London: Faber and Faber, 1939)

Jarman, D., *Derek Jarman's Garden* (London: Thames and Hudson, 1995)

Johnson, L., *The Gardener's Manifesto: Changing the world and creating beauty one garden at a time* (Toronto: Penguin, 2002)

Kelsey, J., *The New Zealand Experiment* (Auckland: Auckland University Press and Bridget Williams Books, 1995)

Kennelly, A., *The Home Vegetable Garden* (Wellington: Department of Agriculture, 1953)

King, M., *The Penguin History of New Zealand* (Auckland: Penguin, 2003)

Lawrence, J., *Gardening Tales* (Wellington: Millward Press, 1981)

Leach, H., *1,000 Years of Gardening in New Zealand* (Wellington: Reed, 1984)

Lee, L., and Lam, R., *Sons of the Soil: Chinese market gardeners in New Zealand* (Pukekohe: Dominion Federation for New Zealand Chinese Commercial Growers, 2012)

Low, B., *Land and People in Christian Order* (Christchurch: Presbyterian Bookroom, 1943)

Mackay, M., *A Compendium of Official Documents Relative to Native Affairs in the South Island*, vol. 2 (Wellington: Government of New Zealand, 1872)

Mann, R., 'Your garden', in Geoff Bryant (ed.), *The Ultimate New Zealand Gardening Book* (North Shore City: David Bateman, 1995)

Massingham, H. (ed.), *England and the Farmer: A symposium* (London: B.T. Batsford, 1941)

Matthews, J., *Matthews on Gardening: A practical approach to the growing of fruits, flowers and vegetables* (Wellington: Reed, 1960)

McCaskill, L., *Hold this Land: A history of soil conservation in New Zealand* (Wellington: A.H. & A.W. Reed, 1973)

McPherson, J., *The Complete New Zealand Gardener* (Christchurch: Whitcombe & Tombs, 1968)

McPherson, J., *Whitcombe's Complete New Zealand Gardener* (Christchurch: Whitcombe & Tombs, 1943)

Mein Smith, P., *A Concise History of New Zealand* (Melbourne: Cambridge University Press, 2005)

Mein Smith, P., *Mothers and King Baby: Infant survival and welfare in an imperial world: Australia 1880–1950* (London: Macmillan Press, 1997)

Mitchell, A., *The Half-gallon Quarter-acre Pavlova Paradise* (Christchurch: Whitcombe & Tombs, 1972)

Moon, P., *The Struggle for Tāmaki Makaurau: The Māori occupation of Auckland to 1820* (Auckland: David Ling, 2007)

Moore, M., *Children of the Poor* (Christchurch: Canterbury University Press, 1996)

Murphy, M., *Handbook for Gardening in New Zealand with a Section on Poultry and Bee-Keeping* (Christchurch: Whitcombe & Tombs, 1895)

New Zealand Federation of University Women, Canterbury Branch, *St Albans: Swamp to Suburbs. An informal history* (Christchurch: New Zealand Federation of University Women, Canterbury Branch, St Albans, 1989)

New Zealand Federation of University Women, *Sydenham: The model borough of old Christchurch. An Informal History* (Christchurch: New Zealand Federation of University Women, Canterbury Branch, 1977)

Ogilvie, G., *The Port Hills of Christchurch* (Wellington: A.H. and A.W. Reed, 1978)

Paul, R., *Letters from Canterbury* (London: Rivington, 1857)

Pritchard, D., *Vegetable Growing in the Home Garden* (Wellington: Dig for Victory Campaign, 1944)

Rice, G., *Heaton Rhodes of Otahuna: The illustrated biography* (Christchurch: Canterbury University Press, 2001)

Roberts, E., *Remembered: The life and work of Ben Roberts, MP* (Masterton: Masterton Publishing, 1965)

Rodale, J., *The Organic Front* (Pennsylvania: Rodale Press, 1948)

Ross, K., *Going Bush: New Zealanders and nature in the twentieth century* (Auckland: Auckland University Press, 2008)

Rutherford, J. (ed.), *The Founding of New Zealand* (Wellington: A.H. and A.W. Reed, 1940)

Shand, Y., *Crime Against the Land* (Lower Hutt: Hutt Printing and Publishing Co., 1941)

Simpson, P., *Dancing Leaves: The story of New Zealand's cabbage tree* (Christchurch: Canterbury University Press, 2000)

Sinclair, K., *A History of New Zealand* (rev. edn) (Auckland: Penguin, 2000)

Sissons, J. et al., *Ngā Pūriri o Taiamai* (Auckland: Reed, 2001)

Smith, B., *European Vision and the South Pacific* (New Haven: Yale University Press, 1985)

Stone, R., *From Tamaki-Makau-Rau to Auckland* (Auckland: Auckland University Press, 2001)

Strongman, T., *The Gardens of Canterbury* (Wellington: A.H. & A.W. Reed, 1984)

Strongman, T., *City Beautiful: The first 100 years of the Christchurch Beautifying Association* (Christchurch: Clerestory Press, 1999)

Tannock, D., *Manual of Gardening in New Zealand* (Auckland: Whitcombe & Tombs, 1914)

Taylor, N., *The New Zealand People at War: The home front*, vol. 1 (Wellington: Department of Internal Affairs, 1986)

Taylor, N., *The New Zealand People at War: The home front*, vol. 2 (Wellington: Department of Internal Affairs, 1986)

Tipples, R., *Colonial Landscape Gardener: Alfred Buxton of Christchurch, New Zealand, 1872–1950* (Lincoln: Lincoln University Department of Horticulture and Landscape, 1989)

Tunnicliff, S. (ed.), *The Selected Letters of Mary Hobhouse* (Wellington: Daphne Brasell Associates Press, 1992)

Wakefield, E., *Adventure in New Zealand from 1839 to 1844*, vols. 1 and 2 (London: John Murray, 1845)

Ward, L., *Early Wellington* (Christchurch: Capper Press, 1975)

Waugh, T., *Home Garden Fertilisers and How to Use Them* (Wellington: Reed, 1945)

White, J., 'Just Gardening' (Wellington: Social Justice Commission of the Anglican Church in Aotearoa, New Zealand and Polynesia, 2009)

Wihongi, H., *Reintroduction of Early Kumara Varieties* (Auckland: Puhaorangi Māori Culture & Resource Centre, 2010)

Wylie, C., *A Garden at My Door* (Christchurch: Caxton, 1964)

Yates Garden Guide (Auckland: Arthur Yates and Co., 1957)

Chapters in edited books

Beattie, J., 'The Empire of the Rhododendron: Reorienting New Zealand garden history', in Eric Pawson and Tom Brooking (eds), *Making a New Land. Environmental histories of New Zealand* (Dunedin: Otago University Press, 2013)

Beattie, J., 'Thomas McDonnell's Opium: Circulating, plants, patronage and power in Britain, China and New Zealand, 1830s–1850s', in Sarah Burke Cahalan and Yota Basaki (eds), *The Botany of Empire in the Long Eighteenth Century* (Washington, DC: Dumbarton Oaks/Harvard University Press, 2017)

Bulmer, S., 'Nga Mara – Traditional Māori Gardens', in M. Bradbury (ed.), *A History of the Garden in New Zealand* (Auckland: Penguin, 1995)

Conan, M., 'From Vernacular Gardening to a Social Anthropology of Gardening', in M. Conan (ed.), *Perspectives on Garden Histories* (Washington, DC: Dumbarton Oaks, 1999)

Davison, G., 'The City-Bred Child and Urban Reform in Melbourne 1900–1940', in Peter Williams (ed.), *Social Process and the City* (Sydney: Allen and Unwin, 1983)

Feuchtwanger, W., 'Gorst, Sir John Eldon', in H. Matthew and B. Harrison (eds), *Oxford Dictionary of National Biography: From the earliest times to the year 2000*, vol. 22 (Oxford: Oxford University Press, 2004)

Giles, David Boarder, 'Fear of a Free Lunch: Markets, Publics and the Would-be Gift', in A. Dundon and R. Vokes (eds), *Shifting States: New Perspectives on Security, Infrastructure and Political Affect* (London: Bloomsbury, forthcoming)

Harris, R., 'As it was: Early Māori and European settlement', in S. Owen (ed.), *The Estuary: Where our rivers meet the sea* (Christchurch: Parks Unit, Christchurch City Council, 1992)

Leach, H., 'Exotic Natives and Contrived Wild Gardens: The twentieth century home garden', in Eric Pawson and Tom Brooking (eds), *Environmental Histories of New Zealand* (Melbourne: Oxford University Press, 2002)

Morris, M., 'Cabbage Wilson's Garden: Overwriting indigenous space in Christchurch, New Zealand, 1850–1856', in Alex Gerbaz and Robyn Mayes (eds), *Palimpsests: Transforming communities* (Perth: Curtin University, 2005)

Pawson, E., 'Confronting Nature', in Graeme Dunstall and John Cookson (eds), *Southern Capital: Christchurch: Towards a city biography 1850–2000* (Christchurch: Canterbury University Press, 2000)

Raine, K., and Adam, J., 'The Settlers' Gardens', in Matthew Bradbury (ed.), *A History of the Garden in New Zealand* (Auckland: Viking, 1995)

Raine, K., 'Domesticating the Land: Colonial women's gardening', in Bronwyn Dalley and Bronwyn Labrum (eds), *Fragments: New Zealand cultural and social history* (Auckland: Auckland University Press, 2000)

Raine, K., 'The First European Gardens', in Matthew Bradbury (ed.), *A History of the Garden in New Zealand* (Auckland: Viking, 1995)

Roche, M., 'The State as Conservationist, 1920–1960: "Wise use" of forests, lands, and water', in Eric Pawson and Tom Brooking (eds), *Environmental Histories of New Zealand* (Melbourne: Oxford University Press, 2002)

Star, P., and Lochhead, L., 'Children of the Burnt Bush: New Zealanders and the indigenous remnant, 1880–1930', in Eric Pawson and Tom Brooking (eds), *Making a New Land: Environmental histories of New Zealand* (Dunedin: Otago University Press, 2013)

Vercoe, W., 'A Self-Sufficient Māoridom', in Witi Ihimaera (ed.), *Vision Aotearoa: Kaupapa New Zealand* (Wellington: Bridget Williams Books, 1994)

Journal articles

Adam, J., 'The Leisured Classes of Ponsonby: Gardens and parks', *The New Zealand Map Society Journal*, no. 13, 2000

Barton, G., 'Empire Forestry and the Origins of Environmentalism', *Journal of Historical Geography*, vol. 27, no. 4, 2001

Barton, G., 'Sir Albert Howard and the Forestry Roots of the Organic Farming Movement', *Agricultural History*, vol. 75, no. 2, 2001

Beattie, J., 'Environmental Anxiety in New Zealand, 1840–1941: Climate change, soil erosion, sand drift, flooding and forest conservation', *Environment & History*, vol. 9, no. 4, 2003

Beattie, J., and Stenhouse, J., 'Empire, Environment and Religion: God and the natural world in nineteenth-century New Zealand', *Environment & Society*, 13, 2007

Beattie, J., Heinzen, J., and Adam, J., 'Japanese gardens and plants in New Zealand, 1850–1950: Transculturation and transmission', *Studies in the History of Gardens and Designed Landscapes*, vol. 28, no. 2, 2008, 219–36

Beattie, J., 'Scientific Agriculture, Health and Gardening: Japan, New Zealand and Bella and Frederic Truby King', *New Zealand Journal of Asian Studies*, vol. 16, no. 2, December 2014

Beattie, J., and Boileau, J., '"Cultivated with great carefulness": Chinese market gardening, urban food supplies and public health in Australasia, 1860s–1950s', forthcoming, 2020.

Birdsall, R., 'Commercial Aspect and Future Possibilities of Rotenone', *Industrial & Engineering Chemistry*, vol. 25, June 1933

Bulmer, Susan, 'Sources for the Archaeology of the Maaori Settlement of the Taamaki Volcanic District' (Wellington: Department of Conservation, 1994), 41–42

Clemens, J., 'Clianthus: Veiled in mystery', *New Zealand Garden Journal*, vol. 5, no. 1, June 2002, 4–5: www.rnzih.org.nz/pages/Clianthus.htm

Challenger, S., 'Horticulture in Canterbury: Recent activities and developments', *New Zealand Gardener*, vol. 15, no. 11, 1959

Challenger, S., 'Studies on Pioneer Canterbury Nurserymen (1) William Wilson', *Annual Journal of the Royal New Zealand Institute*, no. 6, 1978

Challenger, S., 'Pioneer Nurserymen of Canterbury, New Zealand (1859-65)', *Garden History*, vol. 7, no. 1, 1979

Cooper, R., 'The Banks Lecture: Early Auckland gardens', *Journal of the Royal NZ Institute of Horticulture*, 1971

Keogh, L., 'The Wardian Case: Environmental histories of a box for moving plants', in *Environment and History*, vol. 25, no. 2 (Winwick: White Horse Press, 2019)

Holmes, K , '"I have built up a little garden": The vernacular garden, national identity and a sense of place', *Studies in the History of Gardens and Designed Landscapes*, vol. 21, no. 2, 2001

Lee, L., and Lam, R., '陈达枝 Chan Dah Chee (1851–1930): Pioneer Chinese market gardener and Auckland businessman', *Environment and Nature in New Zealand*, vol. 6, no. 1, July 2011

Morris, M., 'A Celestial Place: Hill gardening in a colonial garden city', *Thesis Eleven*, no. 92, February 2008

Prebble, M. et al., 'Early Tropical Crop Production in Marginal Subtropical and Temperate Polynesia', *Proceedings of the National Academy of Sciences of the United States of America*, vol. 116, no. 18, 30 April 2019

Roche, M., '"Wilderness to Orchard": The export apple industry in Nelson, New Zealand 1908–1940', *Environment and History*, vol. 9, no. 4, 2003

Shadbolt, T., 'A Personal Comment on the Urban Environment in New Zealand', *The Landscape: Journal of the New Zealand Institute of Landscape Architects Inc*, no. 21, April 1984

Watts, M. 'Parkinson's Disease, Pesticides and Derris Dust', *Organic NZ*, Jan/Feb 2001: www.organicnz.org/organic-nz-magazine/1138/parkinsons-disease-pesticides-and-derris-dust/

Wood, V., 'Appraising Soil Fertility in Early Colonial New Zealand: The "biometric fallacy" and beyond', *Environment and History*, vol. 9, no. 4, 2003

Technical reports

Anderson, A., 'Mahinga Kai: Evidence for the Ngai Tahu claim to the Waitangi Tribunal' (Tuahiwi: Ngai Tahu, 1988)

Furey, L., 'Maori Gardening: An archaeological perspective' (Wellington: Science and Technical Publishing, Department of Conservation, 2006)

MacGibbon, L., and M. Morton, 'It's a lot of little things happening that will make the difference: A community needs analysis of the Shirley area' (Christchurch: Shirley/ Papanui Community Board, 2001)

St Albans Working Party, 'Final Report to Council and Community', 13 January 2009

Tau, T.M., 'Mahinga Kai: Tuahiwi evidence for the Ngai Tahu claim before the Waitangi Tribunal' (Tuahiwi: Ngai Tahu, 1988)

Tau, T.M., 'Cultural Report on the Southwest Area Plan for the Christchurch City Council' (Christchurch: Christchurch City Council, 2005)

Tau, T.M., Goodall, A., Palmer, D., and Tau, R., 'Te Whakatau Kaupapa: Ngai Tahu resource management strategy for the Canterbury region' (Wellington: Aoraki Press, 1990)

Te Urewera, WAI 894 (Wellington: Waitangi Tribunal, 2009)

Wellington Tenths Trust, GIS Map Book (Wellington: Borcoda Print, 2004)

Theses

Acres, B., 'Opportunities for Food Systems Planning in New Zealand' (MPlan thesis, University of Otago, 2010)

Bainbridge, A., 'Birth, Death, and Marriage in the Garden: Canterbury colonial women gardeners, 1850–1914' (MA thesis, University of Waikato, 2015)

Bishop, J., 'The Role of Medicinal Plants in New Zealand's Settler Medical Culture, 1850s–1920s' (PhD thesis, University of Waikato, 2014)

Gush, N., 'Beauty of Health: Cora Wilding and the Sunlight League' (MA thesis, University of Canterbury, 2003)

Hammetter, J., 'Gardening and Meaningful Lives: An ethnography of Milwaukee-area residential vernacular gardens' (PhD thesis, Northwestern University, 2002)

Heine, J., 'Colonial Anxieties and the Construction of Identities: The employment of Maori women in Chinese market gardens, Auckland, 1929' (MA thesis, University of Waikato, 2006)

Jordan, K., 'Golden Weather Gardening: New Zealand home sections, 1945–1970' (MA thesis, University of Auckland, 2010)

Loeffler, F., 'Pteridomania: A visual history of the fern in New Zealand' (MA thesis, Victoria University of Wellington, 2006)

Morris, M., 'A History of Christchurch Home Gardening from Colonisation to the Queen's Visit: Gardening culture in a particular society and environment' (PhD thesis, University of Canterbury, 2006)

Oakley, P., 'The Handling of Depression Problems in Christchurch, 1928–35: A social study' (MA thesis, University of Canterbury, 1953)

Watson, J., 'Crisis and Change: Economic crisis and technological change between the world wars, with special reference to Christchurch, 1926–36' (PhD thesis, University of Canterbury, 1984)

Wilson, S., 'The Aims and Ideology of Cora Wilding and the Sunlight League 1930–1936' (MA essay, University of Canterbury, 1980)

Websites

'About the Localising Food Project': www.localisingfood.com/index.php/the-project

'Agent Orange': www.publichealth.va.gov/exposures/agentorange

'An act to prevent the introduction and to provide for the eradication of diseases affecting orchards and gardens': www.enzs.auckland.ac.nz/docs/1896/1896A45.pdf

'Climate Smart Strategy 2010–2025': https://ccc.govt.nz/assets/Documents/The-Council/Plans-Strategies-Policies-Bylaws/Strategies/ClimateSmartStrategy2010-2025.pdf

Food Forest NZ: http://foodforest.co.nz/about/

'Food Resilience Policy': www.ccc.govt.nz/the-council/plans-strategies-policies-and-bylaws/policies/sustainability-policies/food-resilience-policy/

'Historical Use of Lead Arsenate Insecticides, Resulting Soil Contamination and Implications for Soil Remediation: http://soils.tfrec.wsu.edu/leadhistory.htm

'Low Levels of Dioxin in Residential Soils at Paritutu in New Plymouth': www.mfe.govt.nz/publications/hazardous/taranaki-dioxin-report-sep02/introduction-sep02.pdf

McGregors Seed Catalogue: www.mcgregors.co.nz/products/showprod.php?sec=3&cat=NZNA&prod=M4400

'Native Plants in Lincoln Township: Ecological Baseline Survey of Residential Properties': www.lincolnenvirotown.org.nz/docs/Native-Plants-in-Lincoln-Township.pdf

'Report on the Hagley/Ferrymead Community Board to the Council of 30 January 2002': www1.ccc.govnz/Council/Proceedings/2002/February/HagleyFerrymead/

'Ti Kouka Whenua': http://library.christchurch.org.nz/TiKoukaWhenua/

'Tuberculosis and the Māori People': www.nzonscreen.com/title/tuberculosis-and-the-Māori-people-1952

'Vegetable Gardens and Other Domestic Sources of Soil Contamination in Residential Property': www.wasteminz.org.nz/conference/conferencepapers2006/Brian%20Maunder%20-%20Vege%20Gardens.pdf

'Weed Management Policy': www.aucklandcity.govt.nz/council/documents/resources/weed/section3-5.asp

www.gardentotable.org.nz
www.inchem.org/documents/icsc/icsc/eics0765.htm
www.mfe.govt.nz/publications/hazardous/risks-former-sheep-dip-sites-nov06/html/
    page14.html
www.nzhistory.net.nz/people/henare-tomoana
www.transitiontowns.org.nz/node/341

Index